Tools of the Trade
Equipping the Canadian Army

edited by
Doug Knight

additional text by
Clive M. Law

Service Publications
Box 33071 Ottawa, Ontario,
Canada K2C 3Y9
www.servicepub.com

© Service Publications

This edition first published in Canada,
April 2005 by
Service Publications,
PO Box 33071
Ottawa, Ontario K2C 3Y9
Ph - 613-820-7350
Fx - 613-820-1288
sales@servicepub.com
www.servicepub.com

ISBN 1-894581-23-7

Printed and bound in Canada
Cover and book design by Clive M. Law

Weapons of War Series editor-in-chief - Steve Guthrie

*Cover photo; Crew of a triple 20mm Polsten anti-aircraft gun searches the
skies above England while undertaking training. DND Photo Z-1082-2.*

*Title page; "Netting Wireless Sets" 1 Canadian Armoured Car Regiment,
(RCD), Hove, Sussex, 12 April 1943, pencil study of a Mark III Canadian Scout
Car (Ford) by E.J Hughes.*

Acknowledgements

The publisher wishes to thank Steve Guthrie, Doug Knight,
Mark W Tonner and Barry Beldam for their willingness to read
through the drafts and to offer their suggestions and encouragement.
Dana Nield and Barry Beldam for providing important photos. The
staff at the National Archives, Research Services Division, National
Photograph Collection for their always-courteous help, with a special
mention to Jean Matheson.

TOOLS OF THE TRADE
Equipping The Canadian Army

Introduction

Throughout the Second World War the Canadian Military Headquarters (CMHQ) compiled a monthly report of the status of equipment on issue and due to units of the Canadian Army Overseas. These monthly reports formed the basis for four (almost) annual reports in the CMHQ reports series overseen by Colonel C.P Stacey, doyen of Canadian military history, and written by Stacey, Captain GTJ Barrett, Captain JEA Crake and Lieutenant JR Martin. The text of these reports makes for fascinating reading, even in their militarily abbreviated lexicon liberally laced with obscure file references. It was felt that if these four reports could be merged and the actual text cleaned up so as to permit a more flowing narrative, then a clearer picture of both the incredible difficulties and the heartening successes of supplying an ill-prepared peacetime army for the serious business of war could be presented to students of the Second World War.

If it took four war-time historians to create the original reports it took another four modern historians to make this treatise possible. I gratefully acknowledge the invaluable assistance of Doug Knight, Steve Guthrie, Mark W. Tonner and Barry Beldam. Their willingness to read through early drafts, make valuable suggestions, edit large tracts of the text and suggest captions, makes this work as much theirs as the original authors'.

In addition to the edited text, additional information on unusual items, as well as experimental projects, has been added. Finally, a wide selection of photographs, illustrating the equipment in question, has been included. These will certainly assist novices in identifying the various small arms, weapons systems and vehicles.

Ultimately, the intent of "Tools of the Trade" is to provide an overview of the types of equipment encountered by the Combat Arms of the Canadian Army and not necessarily a definitive history of the acquisition and use of any one particular model. For more in-depth information on any single weapon or vehicle, readers are encouraged to read the specific titles in the related "Weapons of War" series.

Clive M. Law
April 2005
Ottawa

A contemporary photo of a Canadian infantryman. Canadian policy was to supply a number of Canadian-made goods to Canadian soldiers. This soldier is most likely wearing a Canadian-made Mark II steel helmet, Pattern 1937 Web Equipment, a No. 4 Lee-Enfield rifle and a Pattern 1937 battledress uniform. Photo courtesy Department of National Defence - ZK-717

Part 1

The 1935 experimental armoured cars were an early attempt at developing a Canadian armoured vehicle. Both Ford (left) and General Motors (right) were encouraged to produce these models at almost no cost to the Canadian government. Both companies declined to continue development without more financial involvement from the government. Photo courtesy Royal Canadian Dragoons Museum.

The Beginning - 1939-1940

When the 1st Canadian Infantry Division (then termed the 1st Division, Canadian Active Service Force) arrived in Britain in December 1939, it was mostly armed with First World War weapons. With few exceptions Canadian sources of modern weapons did not yet exist and Britain's factories had been working to capacity to fill British Government orders. The result was that the infantry units in the Division crossed the ocean armed with the .303-inch Short Magazine Lee Enfield (SMLE) rifle of World War 1 fame and the Lewis light machine gun instead of the more modern Bren. The only artillery weapon available in any quantity in Canada was the 18-pounder gun, a weapon inferior to the new 25-pounder gun-howitzer. Due to the need to retain the 18-pounders in Canada for training purposes, the field artillery brigades in the Division (they were reorganised into field regiments on their arrival) arrived in Britain without guns. After being provided for a time with four hybrid 18/25 pounders per regiment (18-pounders that had been converted to 25-pounders guns), they were obliged to relinquish these and were re-equipped with the obsolescent 18-pounders. The 1st Canadian Medium Regiment, which arrived in the United Kingdom in February 1940, brought their aging 6-inch howitzers from Canada.

Through the winter and the spring of 1940 the process of equipping the 1st Canadian Infantry Division and its ancillary units from British sources went on gradually. The General Report for the 1st Canadian Infantry Division for the week ending 24 February 1940 observed that, although the British War Office had given the Division preferential treatment in the provision of training equipment, the limited quantities, particularly of vehicles, was hampering progress. Two weeks later deliveries of Bren guns, sufficient to issue 40 per rifle battalion, and of Boys Anti-tank rifles to complete each unit establishment, were expected to begin shortly. Universal Carriers (small, lightly-armoured tracked vehicles) began to arrive in March, and at the end of the month the 1st Canadian Anti-tank Regiment received six 2-pounder anti-tank guns, giving them a total of eight against a war establishment of 48 guns. A 'war establish-

Early in the war Canada obtained 250 US-made M-17 light tanks (based on the Renault FT-17). The tanks were of use solely as training vehicles and were disposed of to heavy industry as tractors before war's end.

ment' is a high-level document that lays out a unit's organisation and role, and its entitlement to personnel (by rank and trade), weapons, and major equipments (trucks, tanks, artillery, etc.) A scale of issue is a detailed entitlement for equipment (and there are many types of scales). For example, a mortar platoon may have an establishment for eight medium mortars, each of which has a scale that lists the tools and accessories that are issued with each mortar.

By the spring of 1940, the shortage of Mechanized Transport (MT) - trucks, motorcycles, etc. - was a regular complaint. The situation improved in May with the deliveries of Canadian vehicles. On 18 May 1940, the division reported that eight 15-hundredweight (15-cwt) trucks had been received from the Canadian vehicle assembly plant in Southampton (as a unit of measure, hundredweight (cwt) was 112 pounds (50.9 kilograms)). To save shipping space, Canadian vehicles were shipped from Canada in a "knocked-down state" and were assembled by British workers at the Canadian Mechanization Depot. The depot at Southampton was later destroyed in an air attack and was moved to Slough. The following week's report stated that a further 344 Canadian 15-cwts had been received. The heavier types of Canadian-manufactured vehicles arrived more slowly. By 6 July 1940, Canadian 30-cwt and 3-ton lorries (trucks) were not yet available for issue.

Coincident with the German attack in May 1940, there was a significant improvement in the equipment of the Canadian force. By 1 June the 1st Anti-tank Regiment had 42 guns (of their establishment of 48), the divisional artillery regiments 48 (out of 72), the two Army field regiments 24 (out of 48) and the medium regiment was completely equipped. Approximately 700 vehicles of all types had been taken over as an emergency issue from British stocks.

In June 1940, the 1st Canadian Infantry Division was ordered to France, and its units were issued their full, or nearly full, entitlement of materiel. The report for the week ending 8 June 1940 noted that the division's equipment was rapidly approaching completion at Aldershot, and the non-divisional units were not far behind.

In an attempt to relieve pressure on the British Expeditionary Force in the Dunkirk region, the 1st Canadian Infantry Brigade Group actually landed in France, but was withdrawn almost immediately. It was ordered to destroy most of its transport and other heavy equipment before re-embarking. Even 1st Canadian Field Regiment, Royal Canadian Horse Artillery, which saved its guns, lost much of its other equipment. The consequence was, that when the Canadian force moved to the Oxford area late in June, the units which had landed in France

The Canadian Militia purchased six Carden-Loyd carriers in 1931 and another six in 1932. The carriers entered service with the three Permanent Force infantry regiments in 1933, each regiment receiving four carriers each. Photo courtesy Royal Canadian Dragoons Museum

were obliged to remain at Aldershot for re-equipping. This resulted in the Division being one brigade group short. During the week ending 20 July, the 1st Canadian Infantry Brigade and 1st Canadian Field Company rejoined the Division, but the 1st Canadian Field Regiment, 4th Canadian Field Ambulance, and 1st Canadian Medium Regiment were still not operational, due mostly to deficiencies in MT. In the next few weeks the worst shortages were filled, and the 1st Canadian Infantry Division, and ancillary troops, were prepared for battle.

By this time units of the 2nd Canadian Infantry Division were beginning to arrive in the United Kingdom, and the problem of equipping them proved very difficult. On 20 August 1940, Lieutenant-General A.G.L. McNaughton, then General Officer Commanding 7 British Corps (which included the Canadian forces) wrote a letter to the Senior Officer, Canadian Military Headquarters (CMHQ), expressing his concern over the matter.

CMHQ was the administrative headquarters for the Canadian Army in Britain, and dealt mainly with the British War Office. The field force was progressively increased from a division to a corps to an army as successive Canadian formations arrived in Britain. McNaughton was the Canadian commander at each level until his resignation as army commander in late 1943. On Christmas Day 1940, 7 Corps became 1 Canadian Corps. A Division was a relatively fixed organisation. Normally, two or more divisions were organised into a Corps, which had a number of independent units (artillery, transport, etc.) to support the divisions. These ancillary units were known as "Corps troops" and their counterparts at Army level as "Army troops".

McNaughton's letter dealt with the supply of equipment both for 2nd Canadian Infantry Division and for the Corps troops. He noted that it appeared to be impossible to obtain either the necessary equipment, or any assurance as to when it would be made available. This appears to be due to the apparent division of responsibility for the allocation of equipment between the War Office and General Headquarters of the British Home Forces. He made special mention of the Corps engineer equipment. Another serious matter, particularly in view of the air menace, was the absolute lack of light anti-aircraft guns in the Canadian Forces or in the British units associated with the Canadians in 7 Corps. McNaughton's letter went on to state *"We are*

A 4.5-inch howitzer Mark 1R, still without pneumatic tires. Judging from the uniforms, the time period is in the 1930's. Photo courtesy Terry Hunter

dependent entirely on small arms fire for local protection against hostile aircraft and already casualties are being experienced. I feel that the time has arrived when we must insist on positive information and, if no definite assurance of supply of equipment can be given by the War Office, we should consider advising Canada against the despatch of any additional troops to this country until they can be equipped properly". He requested CMHQ to arrange a meeting with representatives of the War Office who could speak with authority on this matter.

The consequence was a meeting at the War Office on 30 August 1940, attended on behalf of Canada by Generals McNaughton and Montague, Brigadiers Loggie, Turner, and Pope and Colonel McQueen, and on behalf of the War Office by Major-Generals Carr, Morris, and Hicks. At this meeting McNaughton asked for information to enable him to advise the Canadian government regarding the despatch of more troops. He pointed out that certain Corps troops present in the United Kingdom (especially engineer and medical units) were not fully equipped. Even worse, nothing had yet been provided, even for training purposes, for 2nd Canadian Infantry Division units which had been in England for nearly a month.

In reply, General Carr explained that equipment had to be placed where it was most needed and that large consignments to the Middle East were necessary. To meet these requirements it had been necessary to stop issues to units at home. No Bofors (40-mm light anti-aircraft guns) were available in England except for units assigned to Air Defence Great Britain (ADGB). All the others were going to Haifa and Alexandria in Egypt.

The Canadian requirements were discussed in detail, and the following points emerged. No Corps engineering units in the United Kingdom, apart from those in 7 Corps, had more than 25 per cent of their war establishment equipment. The whole output of Boys anti-tank rifles was going to the Middle East until 15 September. Thereafter, two brigades (for Iceland and West Africa) had to be equipped before 2nd Canadian Infantry Division could begin to receive its 25 percent training scale. Field guns would be still later. Available stocks of tracer small arms ammunition (SAA) were very small. Some 83 million rounds of SAA had been left in France, but the situation was improving and about two million rounds a week could be provided to field formations for training purposes. Canadian troops would receive their quota through British Home Forces.

General McNaughton then asked whether, in the circumstances, additional formations should be sent from Canada. He outlined the proposed size of the Canadian overseas force, but

A 4.5-inch howitzer Mark 1PA. The carriage has been modified for pneumatic tires and an improved recoil system probably about 1937. The gun was replaced by the 25-pounder, but remained in service for training during most of the war. Milart photo archives.

observed that the despatch of troops who could not be equipped made no useful contribution. The British officers replied that 24 Divisions were then in England and in various stages of being equipped and trained. General McNaughton remarked that, in view of the information given to him, he had grave doubts as to whether additional formations should be sent from Canada in the spring of 1941.

It was, of course, true that units with no immediate operational role did not need a complete issue of equipment. Brigadier J.H. MacQueen at CMHQ pointed out that *"had Canadian formations undertaken active operations, they would have been made operationally complete with the most modern equipment available"*. In these circumstances, equipping the 2nd Canadian Infantry Division proceeded very slowly. By November 1940, the Divisional artillery, which had been two months in England, had received only 28 guns, and these were out-dated 75-mm First World War weapons.

In the fall of 1940, the British Army Council decided to reorganise the basic infantry division, and this reorganisation required a greater scale of equipment. Canadian organisations were similar to the British, and CMHQ asked the War Office to explain the ramifications of the reorganisation. On 7 November, Major-General A.E. Nye sent a discouraging reply. He explained that 60 percent of all available equipment was sent overseas, and 40 percent was allocated for troops at home. "Home" included Home Forces, Anti-aircraft Command, British troops in Ireland and Iceland, training establishments, and new units being formed. The letter stated: *"It is the function of the War Office to decide how the 40% available for home will be allocated between the various customers, and you will appreciate that the percentage coming to Home Forces is small, but the Commander-in-Chief is entitled to dispose of the equipment coming to him as he wishes. In fact so little is coming to him that he virtually has no discretion, because there are certain units whose situation demands that they must have what equipment is available and in practice practically everything has been mortgaged for a few months ahead, and of course there are large deficiencies everywhere."*

As for the Canadian forces, General Nye pointed out that the equipment requested for them was precisely those items in which the supply was smallest, and added, *"The 1st Canadian Division is a mobilized formation, whilst a great majority of divisions in this country are woefully deficient of equipment. From these factors, I think it is reasonable to deduce*

Canada obtained a small quantity of Bantam 4x4, Scout Car, Mark II. These were the forerunner of the ubiquitous Jeep. Photo courtesy Ford of Canada.

that the prospect of the 1st Canadian Division (or indeed any other division) being brought up to the higher standard of equipment due to the recent decision [to reorganise] within the next six months is precisely nil." This letter led General McNaughton to cable Ottawa requesting that all possible measures be taken to expedite Canadian production. He suggested in particular, that with continued air attacks, it was most important that the Canadian troops have something with which to hit back. He asked that the possibilities of obtaining anti-aircraft guns from the United States be further explored.

Starting Canadian Production

In the Second World War, mechanization and the generally increased complexity of modern equipment had made the problem of equipping an army in the field much more difficult than was the case in the First World War, although it was serious enough even then. Moreover, the British Empire was fighting an enemy who had mobilized its industries for military production some six years before the beginning of hostilities.

This was not the case with Britain or Canada, which began to prepare much later (in Canada, the Bren gun contract, signed on 31 March 1938, was an important exception) and limited their preparations when they did begin. Even after the outbreak of war neither country fully exploited their manufacturing capability until German victories in the spring of 1940 showed the desperate peril of their situation and the full measure of their antagonist's strength. The production of modern weapons by any country whose industry is geared to peacetime tasks is a slow process. After two years of warfare the output of military goods by the British nations was still inadequate to meet the needs of their forces. An armament industry cannot be created overnight, and the Canadian forces would have been much better armed if McNaughton's plan, when he was Chief of the General Staff in the 1930's, for a Dominion Arsenal capable of manufacturing the army's major requirements had been accepted and acted upon by the Canadian government.

Britain's hesitation about letting contracts with Canada should not be overlooked. Until

Vickers Mark VIB Light Tanks undertaking field training at Camp Borden, Ontario - winter 1940. Library & Archives Canada PA129099

Britain left the majority of its modern equipment on the beaches of Dunkerque, it had consistently opposed any major contracts being let in Canada. Part of this was a belief that Canada was not up to the task of building military materiel, but there was also a fear of post-war competition if British technology was shared with Canada. This situation existed with respect to "A" (armoured fighting vehicles) and "B" vehicles (soft-skinned vehicles), optics, aircraft, large-calibre weapons and most other systems. Even Canada's insistence on the ability to produce its own rifle was ignored until the British Government realised that it could not produce rifles or other materiel in sufficient quantities, or within a reasonable time, to supply its own growing forces as well as those of the Commonwealth. In defence of the British hesitance towards Canadian production was the early belief that the war was a "three year conflict", and it was felt within British circles that there was insufficient time to create a Canadian industry. The British also pointed out the technical difficulties that Canada faced in obtaining heavy machine tools from the US to build goods to British specifications. In fact, British hesitancy was not limited to Canada - the British Purchasing Commission, established in New York to administer North American purchases on behalf of the British Ministry of Supply, bought very little during the first six months of the war.

Although Canada was slow to change over to the production of war material, it was able to make an important contribution to the war effort at a comparatively early stage in those areas where her peacetime industries could be directly used. This was most notably the case in the matter of mechanical transport. With a well-equipped and efficient automotive industry, Canada was able to quickly organize production of military trucks on a very large scale. During the war, Canada made an important contribution in this field to the equipment of its own forces and those of the Empire at large. The automotive industry had also been called upon to manufacture Universal Carriers. However, these had less in common with ordinary commercial vehicles and production was slow in getting under way. The preliminary trials of a Canadian-manufactured carrier were held in June 1941.

Thanks only to the fact that a contract was placed over a year before the outbreak of war, the first Bren gun manufactured in Canada was delivered on 23 March 1940. However, this was for a British order, and although the first ten Canadian Brens arrived in the United Kingdom in November 1940, deliveries to Canadian forces remained small. The summary for the week ending 30 August 1941 reported that only 243 were delivered that week, although by 1942 Canada was producing 120,000 per year.

Canadian-made Valentine VI, Infantry Tanks, awaiting delivery of their main armament, parked in rows at Canadian Pacific Railway's Angus Shops in Montreal, 29 January 1942. Library & Archives Canada PA174516

Other war material, which had not been previously manufactured in Canada, entered production very slowly. In 1939, before the outbreak of war, the British Government placed an order for 25-pounder guns with Sorel Industries in Sorel, Quebec, but the first six guns were not handed over until 1 July 1941. However, the first Canadian-manufactured barrel for the Bofors 40-mm anti-aircraft gun was delivered in October 1940, and this was a major achievement. Three-inch mortars entered production in 1940, but the first reference to an issue of them in England was in July 1941. The summary for the first week of September 1941 noted that the first five 2-inch mortars had been delivered. The first Canadian Boys anti-tank rifles were delivered in February 1942.

Once Canadian production got underway, it quickly reached very high levels. This was the case with both Bren guns and Universal Carriers, and was soon be the case with many other weapons. No one can doubt that, in the end, Canada's industrial contribution to victory was very significant. However, the almost painful slowness (in the circumstances, the inevitable slowness) of the process of initiating production was a fundamental lesson that must be stressed. The production of military equipment that Canada needed so desperately can only be rapidly effected in time of war on the basis of facilities established for that purpose in time of peace.

Part of a day's production at Ford of Canada's plant at Windsor, Ontario. In the foreground are a mix of Universal Carriers and Modified Conventional Pattern trucks (MCP). In the backgrond are Canadian Military Pattern (CMP) trucks. Photo courtesy Ford of Canada, AT532

Getting Up to Speed - 1941 and 1942

The basic problem of the Canadian Army Overseas during the first two years of the Second World War, so far as equipment was concerned, was the replacement of 1914-18 pattern weapons. Modern weapons had been developed before the outbreak of war in 1939, but had not been produced in sufficient quantities. After two years of war, the Canadian Army Overseas was still seriously deficient in such basic essentials as modern field guns, light anti-aircraft guns and anti-tank guns. The expansion of British, Canadian, and United States facilities had brought some limited improvement, but certain grave deficiencies remained. A major problem was that this production was geared towards equipment that had been designed long before the war and which was nearly obsolete.

The general situation was clearly indicated in a letter from the Canadian Corps to the CMHQ on 4 January 1941. Referring to the 2nd Canadian Infantry Division, the letter stated *"...advice has been received that only six 25-pounders and no 2-pounder anti-tank guns are being supplied during January. For a division which is complete in personnel and practically complete in MT, this represents a most unsatisfactory position."* Largely on the basis of its peacetime industrial plant, Canada had been able to provide in good time the vehicles needed for 2nd Canadian Infantry Division, but Canada and Britain together had not been able to provide essential elements of its armament. During the early part of 1941, the equipment situation of the division slowly improved. Thanks to Canadian production, the infantry battalions were up to strength in Bren guns before the end of February. At the same time, carriers were being received in considerable numbers, and the artillery situation gradually improved.

It would seem to be accurate to say that the greatly expanded Canadian industrial programme undertaken in 1940 began to bear fruit about midsummer 1941. During June 1941,

Loading a Valentine Infantry tank for export. These tanks were manufactured at Canadian Pacific Railways' Angus Shops in Montreal under British contract but were destined to Russia. Only 30 were retained for training purposes in Canada. Library & Archives Canada PA174520

press reports indicated that production of tanks, both Mark III (Valentine) Infantry and modified M3 Cruiser (Ram) tanks, had begun in Canada. It was understood that examples of both types could be expected in Britain before fall 1941, but neither type was being issued to the Canadian Army Tank Brigade. It had been decided that two battalions of this formation would receive Mark II tanks (Matilda) and the third would receive Mark IV tanks (Churchill).

The somewhat easier situation was reflected in the 3rd Canadian Infantry Division, which arrived in Great Britain during the summer of 1941. This division was much better equipped than its predecessors, and had received field guns in useful numbers. The first two field regiments in the division that arrived in England were complete with 25-pounder guns, and the third regiment, whose arrival was delayed until November 1941, received its 24 guns before the end of that month. By 27 September, the division had received 39 Boys anti-tank rifles, withdrawn from 1st and 2nd Canadian Infantry Divisions, followed a week later by 64 more from British sources. From 27 September to 28 November, the division received 154 carriers from Canadian production. The War Office 'State of Equipment' showed the division as complete by mid-January 1942, with twenty-four 25-pounder guns and thirty-six 2-pounder anti-tank guns. It was also complete in pistols, Bren guns, and Thompson sub-machine guns. There was still a deficiency in Boys anti-tank rifles (87 held against an establishment of 426) and 2-inch mortars (9 out of 40) but the situation was improving. After its arrival in England, in late fall of 1941, the 5th Canadian Armoured Division was quickly fitted out with a large proportion of its equipment.

Units of the 4th Canadian Armoured Division (the last Canadian division to go overseas) arrived in Britain in the summer of 1942. This division's fortunes in matters of equipment differed very greatly from those of the 3rd Canadian Infantry Division and the 5th Canadian Armoured Division, which arrived overseas very incomplete. In contrast, the 4th Canadian Armoured Division brought an average of approximately 75 percent (in some items 100 percent) of its equipment from Canada. This was notably true in the case of signals equipment. The most serious deficiency existed in "A" vehicles. The first issue of Ram tanks took place in November 1942 and was not completed until October 1943.

Assembly of Canadian Military Pattern (CMP) trucks at Slough/Southampton. Most CMP vehicles shipped overseas were 'knocked down' to save shipping space. Re-assembly was performed by British craftsmen under control of the Canadian Mechanization Depot. A special army unit, the 1st Canadian Equipment Assembly Unit, was formed to assist. Library & Archives Canada

Hitting the Stride - 1943

During 1942-43, equipment concerns entered a new phase. The necessity of "making do" with First World War weapons was no longer acceptable. Although factories had produced adequate supplies of 1939-pattern armaments, many of those weapons themselves had become obsolescent as the result of developments during the war. The problem was increasingly one of re-equipping the army with new and more powerful weapons designed from the lessons on the battlefield after 1939. An important example was anti-tank weapons. As late as 1942, Canadian formations were still anxiously awaiting the completion of their establishments of 2-pounder anti-tank guns, but this weapon had already been replaced in the British army by the more formidable 6-pounder, while a newer and heavier gun, the 17-pounder, was entering service. These changes reflected the development of enemy armoured fighting vehicles. The 2-pounder and the Boys anti-tank rifle were adequate against tanks in 1939-40, but more effective weapons were required to stop the heavier Mark IV and Mark VI (Tiger) tanks employed by the Germans in later campaigns. The majority of the tanks available to Lord Gort in the campaign in France in 1940 were armed only with machine-guns (a few having 2-pounders). By 1943, allied tanks were normally armed with the 75-mm gun or the 6-pounder. There were also developments in field gunnery, where changing battlefield conditions brought forth a new class of artillery to support the tanks - the self-propelled gun, which was armoured and mounted on a tank chassis.

Because of these and other developments, Canadian production of war materials had also passed into a new phase. In 1940-42, almost everything was in desperately short supply, and Canada undertook a large production programme involving light and heavy weapons and a great variety of other types of equipment. It had been widely assumed that the Canadian Army Overseas, as well as forces in Canada, would be equipped as much as possible from Canadian sources, and that Canada would also supply large quantities of arms to her allies. In the event, these expectations could not be realized in quite this form, although Canadian production had expanded enormously and certainly had some influence on the course of the war.

19

An Oerlikon 20-mm gun on a naval Mark V pedestal mounting doing guard duty on the roof of Canadian Military Headquarters in London. Most Canadian anti-aircraft units were either under-equipped or were attached to the British anti-aircraft Command (Air Defence Great Britain). Library & Archives Canada e002344118

Equipping Canadian troops was greatly affected by the shortage of sea transport and the importance of using the available tonnage to the best advantage in a war which was being fought in all parts of the globe. This made it necessary to supply the armies in the various theatres, as far as possible, from sources close to the scene of operations, and to develop local manufacturing capacity to the greatest possible extent in those operational areas (for example, India and Australia). In the spring of 1942, two Munitions Assignment Boards were set up to control the global allocation of war materials. The Combined Munitions Assignment Board in Washington dealt with global production assignments, and the London Munitions Assignment Board (LMAB) allocated 'British' war production. The idea was to pool production and assign it to the country that was in the greatest need. Canada, wishing to supply its own forces from its own factories, declined to pool production completely with that of the allies, but was necessarily bound to a large extent by the general Anglo-American policy.

The shipping limitations resulted in the Canadian Army Overseas receiving large quantities of supplies from British sources, while Canadian-made equipment of the same type was shipped directly to the Middle East and other theatres. Also, as allied manufacturing capacity increased, Canada found some of its products becoming a glut on the market, and as a consequence, the manufacturing programme was partially re-oriented. The output of certain items of army equipment (for example, the 6-pounder anti-tank gun and the Valentine tank) was suspended, and a larger proportion of Canada's industrial effort was directed to the construction of ships and aircraft.

Only in the case of certain items, referred to as being of "continuing Canadian supply" was the Canadian Army supplied directly from Canadian sources. Examples were vehicles, certain heavy Engineer earthmoving and tunnelling equipment, and articles of clothing. It must be added that, in addition to these items, the Canadian Army also received great quantities of other articles of Canadian manufacture through LMAB channels under special arrangements. Examples included Canadian Bren and Sten guns and Canadian No. 4 rifles.

Reconnaissance squadrons were organized for the Infantry battalions of the 1st and 2nd Canadian Infantry Brigades in 1940 which utilized motorcycle and sidecar combinations armed with Bren guns. Considerable difficulty in obtaining these motorcycle combinations, most of which were manufactured by Indian, was encountered and these squadrons were disbanded in January 1941. Library & Archives Canada

Apart from the shipping limitations, there were also definite tactical reasons for abandoning any general attempt to equip the Canadian army with Canadian-made material, and in particular with distinctive types of equipment. Had it been certain that all five divisions of the First Canadian Army would be employed together in a single theatre, such a policy would have been practicable. However, the possibility of individual formations being used in scattered theatres had been visualized, and this possibility became reality with the incorporation of one Canadian infantry division, one Canadian army tank brigade, and some Corps troops into the British Eighth Army for the Sicilian campaign. Even though the Canadian component in Italy eventually became a full Corps, and later this Corps rejoined the 1st Canadian Army in Northwest Europe, at the time it appeared improbable that the Canadian Army Overseas would be employed as a single force.

In January 1943, the Canadian Army Overseas was reorganized completely based upon British war establishments. This far-reaching event facilitated cooperation and a more complete integration between formations and units of the British and Canadian Armies. The most important single feature of the reorganization was the adoption of the new British standard for the armoured division. This involved abandoning an organization based upon two armoured brigades and adopting one based upon one armoured brigade, one infantry brigade and a strengthened artillery component. The major part of this reorganization was completed by March 1943.

On the heels of this reorganization came a formal mobilization programme. Mobilization was defined as bringing the unit up to full war establishment in personnel, equipment, clothing and necessaries, and medical equipment. In effect, the unit would have no serious deficiencies that would impair its operational role. Canadian formations were listed in the following order of priority for mobilization: 1st Canadian Infantry Division, 3rd Canadian Infantry Division, 1st Canadian Army Tank Brigade, 5th Canadian Armoured Division, 1st Canadian Corps Troops, 2nd Canadian Infantry Division, 2nd Canadian Corps Troops, First Canadian Army Troops, and finally General Headquarters Troops and Line of Communication Units. 2nd Canadian Infantry Division was placed low in the scale of priority because of the rebuild-

Yet another example of using ingenuity to meet the shortage of anti-aircraft weapons. This mobile plat-form was developed to counter the many low-level reconnaissance missions that Germany flew over Britain. Chevrolet 15-cwt water truck. Library & Archives Canada e002505515

ing required after its heavy losses at Dieppe. 4th Canadian Armoured Division was not included in the original scheme, and was not mobilized until the autumn of 1943.

By the end of 1943, there were still serious equipment problems to be overcome, but they were generally different from those of 1941. There were surpluses of many basic weapons that were formerly in short supply. Other weapons, the lack of which had previously been cause for concern, had become obsolete and had disappeared from Canadian war establishments. These had been replaced by others, which in some cases were not yet available in adequate numbers. The war had become a struggle of technicians, in which scientific effort was directed towards constantly improving the equipment of one's own army, and keeping it better than the equipment of the enemy. The most serious equipment problem in 1943 was the shortage of 20-mm anti-aircraft guns and mountings, but at the time it was anticipated that Canadian production would alleviate this situation.

The Home Stretch - 1944 - 1945

At the beginning of 1944, it became necessary to decide what types of equipment were to be used in operations, having regard to their suitability and the projected state of supply. An important factor in these decisions was British policy. In order to facilitate supply and maintenance, it was desirable to maintain a general uniformity of equipment in each theatre of operations. Splitting the relatively small Canadian forces between two theatres accentuated this factor.

Canadian forces operating with the Allied armies in Italy were supplied through British channels, though responsibility for the provision was Canadian. On arrival in Italy, 5th Canadian Armoured Division took over the vehicles of 7th (Br) Armoured Division. These were in terrible shape and mostly needed replacement. During the year, the Canadian forces in Italy were completely re-equipped with Canadian vehicles.

In Britain, the first five months of 1944 were devoted largely to the process of complet-

This 'Beaverette' is one of nine issued in April 1941 to 4th Reconnaissance Battalion (4th Princess Louise Dragoon Guards), (1st Canadian Division). They were only suitable for training purposes. Library & Archives Canada e002343935

ing the equipment of the Canadian Army for the invasion of Europe. There were two major problems. Not only did the standard types need to be determined and provided in sufficient quantity, but also a considerable amount of special equipment had to be issued to 3rd Canadian Infantry Division and 2nd Canadian Armoured Brigade for the assault landing itself. This special equipment was provided entirely through British sources.

In the last half of 1944 and into 1945, the two main problems were maintaining supplies of equipment in use and developing new weapons and vehicles that experience had shown to be desirable. General policy, decided at CMHQ, occasionally had to be modified by circumstances prevailing in the theatre. This applied to Northwest Europe as well as to Italy. The majority of the "B" vehicles (i.e. soft-skinned vehicles) held by the Canadian forces continued to be Canadian-made, and Canada also made useful contributions in weapons of the standard British pattern. But apart from carriers, few Canadian-made armoured vehicles were adopted for use by the Canadian Army, though some use was found for most of those which had been already delivered. Although the 25-pounder self-propelled guns on the Ram tank chassis (Sexton) were much in demand, the distinctive Canadian 20-mm anti-aircraft gun was not required due mainly to decreased operational demands.

All things considered, Canada's performance in producing all types of war materiel during the Second World War was nothing short of outstanding. Canada not only produced a wide range of British-designed equipment, but also contributed significantly to the design and development of new weapons. It was noteworthy that very little of the Canadian Army's equipment failed to be entirely satisfactory in action.

Handguns

A 40mm Bofors anti-aircraft gun. It's crew of four include a loader, gunner and two men to aim the gun - one for deflection and one for elevation. The insignia and Mark III helmets identify this gun crew as coming from the 3rd Canadian Infantry Division. Photo courtesy Department of National Defence - ZK-1087-4

PART 2

Canadian Provost Corps. Small Arms inspection prior to going on duty. These Military Policemen are armed with the American Smith & Wesson revolver, reflecting the difficulty in obtaining British small arms. Library & Archives Canada e002055517

As early as 1940 it had been determined that Canadian personnel would carry the .380-inch pistol, rather than the heavier .455-inch of the previous war. Supply in quantity of the former did not appear until June 1942 and many were purchased from the US. In that month a substantial number of .380-inch revolvers were issued, and the .455-inch, previously issued as a temporary measure, began to be withdrawn proportionately. By the end of 1943 all .455-inch revolvers had been replaced by the .380-inch.

The principal models of revolvers on issue in the Canadian Army were the Smith and Wesson Military & Police (M&P) in .38 calibre and the Pistol, No. 2, in .380-inch. Owing to the limitations of the calibre these revolvers did not prove entirely satisfactory; officers usually preferring to carry a rifle or machine carbine. The scale of issue of these weapons was materially reduced in March 1943 and those pistols deleted from establishments were replaced by Sten machine carbines. In June 1943, with the Bren Light Machine Gun (LMG) having been declared a personal weapon, pistols were *"withdrawn from the firers of single Bren guns"*, so that a further reduction in establishment was effected.

Meanwhile, the John Inglis Company, Ltd., of Toronto, was contracted to supply a semi-automatic pistol to China and to Great Britain under the Mutual Aid Act. This pistol, based on a pre-war model manufactured by the Belgian firm of Fabrique National, was a substantial improvement over the revolver which had been roundly condemned after Dieppe. A comparison of the two pistols was made by the Historical Officer, CMHQ, 19 April 1945: *"At Corps we picked up Major Wrinch and drove with him to 5th Canadian Armoured Division across the high ground north of Arnhem, a sandy tract of pinewoods. En route we stopped to fire our pistols. I had carried a .38 at intervals since 1940 but had never fired it before. We found that of five .38 bullets striking a German metal ammunition box used as a target, two holed it and bounced off, the rest fell inside; but a bullet from Major Wrinch's 9-mm automatic pistol penetrated both sides of two such boxes placed beside each other. The difference in the performance of the two weapons is striking. The .38, however, appears to be quite accurate."*

In October 1944 it was decided *"to replace with Pistols Automatic Browning 9-mm High Power of Canadian manufacture all Pistols Revolver .380 calibre, and (also) Carbines Machine Sten at present issued to tank crews and signals Despatch Riders (DR) of Canadian*

Soldiers of the 1st Canadian Infantry Division embarking for overseas. They wear Pattern 1908 webbing and carry First World War-vintage rifles, the SMLE no.1 Mk. III*. Library & Archives Canada

The Canadian soldier was issued a variety of rifles in the early days of the war. These included Ross, Pattern 1914 and Enfield SMLE No.1 Mark III rifles. By the time of the Normandy invasion all Canadian soldiers overseas were issued with the No.4 rifle (above) - usually one of Canadian manufacture. Library & Archives Canada

Infantrymen of the 1st Canadian Division on exercise. They are shown here with the No.1 rifle equipped with a Cup, Discharger, for launching grenades. These soldiers previously had their bayonets fixed to the rifles and have placed these on the ground rather then replacing them in their scabbards. Library & Archives Canada e002344112

units operating in North West Europe (NWE) Pistols, Revolver, .380 calibre will remain in use with units operating in the AAI [Allied Armies of the Interior theatre of operations (Italy)]"

By the end of December, 1944 First Canadian Army held about three-quarters of the Browning 9-mm pistols needed to cover the new establishment, and sufficient stocks had been shipped to complete the process of replacement. The 1st Canadian Parachute Battalion, under British command, was equipped with the .45 Colt automatic and, on 30 November 1944, they held exactly enough of these to meet their establishment. With the adoption of the Inglis-made pistol by the Canadian Army they also changed to the Browning 9-mm.

Rifles

Canada entered the Second World War with a mix of World War One vintage rifles, mostly the No.1 Mark III* as standard issue and a quantity of Ross Rifles, Mark III as a war reserve. In spite of Canada's many entreaties to Britain, and despite the failure of the Ross Rifle in the

Soldiers armed with the Thompson Machine Carbine are shown here in training in England. The cost, and difficulty in acquiring these, led to the Sten. Library & Archives Canada e002343945

Firing the Sten Mark II sub-machine gun from the standing position. The photo shows the correct method of supporting the weapon which puts the left hand ahead of the ejection port and not gripping the magazine. Library & Archives Canada e002505358

First World War, Canada still had no domestic rifle manufacturing capability. With Britain's inability to supply rifles to Canada's growing army in the early days of the war 20,000 Pattern '17 (US M1917) rifles were obtained from the US. These rifles had limited use due to their non-standard calibre (US Cal. .30-'06) and were relegated to local defense, reserves and training purposes.

In June 1940, ground was turned at Long Branch, Ontario, for the construction of a plant capable of producing 1,000 rifles per week. The rifle chosen for production at this new facility was the newer Rifle, No. 4, Mark I, and later the Mark I*.

By February 1942 the Canadian Army Overseas still had a considerable shortage of rifles

A Gunner of the 4th Anti-Tank Regiment, RCA in England, 1942 shown here with the recently issued Bren Light Machine Gun. British inability to supply the Canadian Corps with effective anti-aircraft weapons led to the extensive use of the Bren gun in this role. Library & Archives Canada

being, at the time, deficient over 16,000. This shortage was mostly in the reinforcement units whose rifle establishment had recently been increased. The field formations, however, were generally up to strength, albeit with the older No. 1 Mark III* rifle. Although there existed some shortages in both the 5th Canadian Armoured Division and the 1st Canadian Army Tank Brigade, by 31 July 1942, these shortages had been eliminated.

The No. 4 rifle, designed to simplify manufacture, began to be available in quantity from British sources late in 1941, and the question was at once considered whether the Canadian Army Overseas should be equipped with this weapon. It was finally decided in July 1942 to equip all armoured formations in the United Kingdom with No. 4 rifles. This re-equipping pro-

In the absence of sufficient anti-aircraft guns the Canadian Corps used twinned Bren guns. This tasking forced the future development of 100-round magazines. Library & Archives Canada

ceeded very satisfactorily and by November 1942, in order to standardize the rifle and thereby to reduce the variety of parts held as spares, the policy was adopted of arming all of First Canadian Army with the No. 4 rifle. Canadian production of this rifle had reached the stage where considerable shipments were being made overseas. During June and July 1943, large quantities were received from both Canadian and British sources and, on 17 July 1943, the weekly progress report from CMHQ noted: *"All field units now equipped with Rifles No. 4 to 100% War Establishment. The Equipment State of the Canadian Army in the United Kingdom shows a considerable surplus in the supply of rifles, including some No. 1 rifles."*

By the end of January 1944, the Canadian Army was completely equipped with the No. 4 Rifle. This rifle continued in use and in good supply. Under Canada's policy of obtaining Canadian-made goods for Canadian soldiers, the main source was Canadian production from Small Arms Limited and, in the second half of the year, reinforcements from Canada began to arrive in the United Kingdom already equipped with No. 4 rifles. An objection to the Mark I rifle was the aperture sight. In July this began to be replaced on all new rifles by new leaf backsights, the Marks 3 and 4. This up-dated rifle model was known as the No. 4 Mark I*. At the beginning of July the replacement of the aperture battle sight by the Mark 3 for No. 4 rifles held by units was authorized. This modification was performed on the basis of operational priority.

As regards the sniper rifle, the supply situation was not so satisfactory, but there was no marked deficiency on the existing scale of issue. In April 1943, units of the Canadian Army with the Allied Armies in Italy recommended an increase in the scale of issue for the weapon, but the supply situation prevented any immediate action being taken. In June, another request was turned down for the same reason. This situation still prevailed in September, when requests for increases were being received from North-West Europe as well. Initially, the

Two soldiers of the 48th Highlanders with a Vickers "K" Machine Gun. This weapon was often mounted in aircraft or used in an anti-aircraft role. Courtesy the 48th Highlanders of Canada Museum.

Canadian Army was supplied with the No. 1 Mark III* (T) but, by June 1943, first issues of the No. 4 Mark I* (T) were made.

During 1944, the 'Rifle, (Light), No. 5' was developed and production started in the United Kingdom. The first rifles were available for issue by 1 December 1944 and were intended for use in the Far East only. The production policy in force was that Canada would continue to produce the No. 4 only, whereas the United Kingdom would shift entirely to the new model. This caused some concern in Canada over prospects for post-war production.

Machine Carbines

Two principal types of machine carbines were used by the Canadian forces during the war. In the Allied Armies in Italy it was the Thompson machine carbine. Supplies, from British sources, were barely adequate to meet the requirements of the field units and to train and equip reinforcements in the United Kingdom proceeding to Italy. This weapon proved quite satisfactory in action.

During the summer of 1941 the Thompson sub-machine-gun, as it was also known, had become available from the United States in increasing numbers. For example, the Corps General Report noted that 147 of these weapons were issued to Corps units during the week ending 2 August 1941 and the remaining deficiencies in this field were small. At that time, the establishment allowed each infantry platoon three, carried by the section commanders.

By February 1942, the Canadian infantry divisions were almost complete in Thompson sub-machine guns. 3rd Canadian Infantry Division reported 546 on charge against an establishment of 598, while the more senior divisons were much closer to completion.

The situation with regard to Thompson sub-machine guns underwent a radical change in August 1942. In that month the establishment for the weapon was changed in the infantry division from 643 to 1,094, and in the armoured division from 1,412 to 1,573. This change in establishment created a new demand for 5,445 weapons. In other words, an establishment which had been almost complete was suddenly converted to one which was 52 percent deficient. It became apparent that the US production facilities of the Thompson were not adequate

Vickers Medium Machine Gun. Used as a support weapon, the Vickers could be used in both a direct fire mode (see the enemy) as well as an indirect fire mode (enemy is out of the line of sight.) Library & Archives Canada e002343944

to meet the increased demands for automatic weapons by the Allied forces. The need for some weapon other than the Thompson was indicated, if shortages were to be met in reasonable time.

British authorities had developed an alternative weapon - the Sten machine carbine. It fired 9-mm ammunition whereas the Thompson used the heavier .45 inch ammunition. The individual soldier could, accordingly, carry much more Sten ammunition. The Sten had many other advantages over the Thompson. It was much lighter in weight, much less complicated, a great deal cheaper to manufacture, and it could readily be produced in large quantities.

Dieppe was the first occasion on which the Canadians used the Sten and, of the 350 issued, many adverse reports were received upon its performance. A particularly detailed and scathing account came from the Cameron Highlanders of Canada; *"Useless for infantry. The gun is not strongly constructed and is too unreliable for assault work. The welding snapped on one gun. A second would not fire. A third jammed on the second magazine. The barrel blew off the fourth."*

By October 1942, no favourable reports had been received on this weapon. Careful investigation, however, considerably reduced the effect of testimony such as that of the Cameron Highlanders of Canada. An officer interviewed a large number of Canadian participants in the Dieppe operation, and concluded that the troops had had insufficient experience with this weapon before using it in action and did not have time to check them or to familiarize themselves with the gun. They should have been cleaned, fired, and all magazines checked before going into action and, from the evidence given, it appears that Stens and magazines were not checked and inspected adequately before issue to units. These facts show that neither the men nor the weapons were given a fair chance at Dieppe.

With respect to the Camerons' report, note should be taken of the evidence of Major (later Lt-Col) A.T. Law, the senior surviving officer of this unit, who reported that, during the training preceding the Dieppe operation, ammunition for practice was not available, and in consequence the only person in the unit who had actually fired a Sten was the Commanding Officer, Lt-Col Gostling. He added that Sten magazines were provided ready-loaded, and the fact that they were improperly loaded was responsible for many of the difficulties encountered. In light

Cameron Highlanders of Ottawa (MG), 3rd Canadian Infantry Division Machine Gun Battalion, taking aim with the Vickers Medium Machine Gun, Normandy, 1944. Library & Archives Canada PA129037

of this evidence, it was considered that the Dieppe experience did not warrant rejecting the Sten.

In October 1942, 1,048 machine carbines were issued to the Canadian Army Overseas. By 30 June 1943, there was still a considerable deficiency. The shortage was not, however, a serious one. It was found chiefly within the Canadian reinforcement units and what minor deficiencies existed in the field formations had been corrected. By 31 December 1943, the shortage of Stens among Canadian units remaining in the United Kingdom amounted to 10,377 (still largely confined to the reinforcement units), and this was somewhat offset by 557 Thompsons.

Small Arms Limited, of Long Branch, Ontario, delivered its first Sten gun, a Mark II, in February 1942. In June 1943, Sten carbines of Canadian manufacture were adopted as standard for the Canadian Army Overseas, although it was recognized that British weapons would still be issued until sufficient Canadian stocks became available.

It was, however, the Thompson machine carbine, not the Sten, with which the 1st Canadian Infantry Division and the 1st Canadian Army Tank Brigade were armed when they proceeded to the Mediterranean theatre. The Sten was not used in the Middle East or Italian campaigns.

In North-West Europe, supplies of the Sten were adequate, and in May 1944, an additional pool of 12 Stens over and above establishment was allowed to each infantry battalion for the use of platoon commanders as desired. By war's end Small Arms Limited had produced 149,905 Stens for use by Canada, China and Great Britain.

Soon after the Canadians went into action in Normandy complaints began to be received concerning the performance of the Sten. These came from the 3rd Canadian Infantry Division and covered a wide variety of problems. In short, the troops had lost confidence in the weapon.

PIAT Gunner, Italy 1943. The 'Projector, Infantry, Anti-Tank' replaced the Boys rifle as the infantryman's defence against enemy armour. The weapon required that the gunner be close to his target to guarantee a hit. Library & Archives Canada PA189919

It was recommended that either the faults be corrected or that the Sten be replaced by the Thompson sub-machine gun. The complaints were investigated by Second British Army, which, at the time, included the 3rd Canadian Infantry Division. They recommended modifications to incorporate a safety catch, better magazine, and improvement of the selection lever. It was considered that *"good maintenance and cleanliness will eliminate troubles. No other*

The Projector, Infantry, Anti-Tank (PIAT) replaced the Boys rifle as the combat soldiers anti-tank defence. Shown here are three soldiers of the Regiment de la Chaudiere in the distinctive "3 Div" helmets. This helmet, the Mark III, was issued to the actual D-Day assault troops. Library & Archives Canada PA132894

formation reports bear out 3 Canadian Infantry Division report." However the unfavourable reports continued. Early in August various substitutes were being suggested. It should be noted that the objections were not unanimous. The South Saskatchewan Regiment, in July, reported the Sten gun to be excellent; *"we killed more Jerries with Stens ... than with any other weapon".*

Apart from the replacement of a few of the Stens by the Browning 9-mm automatic pistol, no change took place in policy with regard to machine carbines in 21st Army Group. In Italy limited issues were made to British forces of the American .45-inch M3 sub-machine gun, though up to the end of November 1944, Canadians had not received any. Canadian reinforcement units in the United Kingdom held a small number of these for training reinforcements destined for Italy.

Sten production in Canada had ceased by the end of 1944. In the United Kingdom it had switched over entirely to the new Mark 5. The Mark 5 incorporated a wooden butt and pistol grip, a foresight similar to that on the rifle, and other improvements. It was designed primarily for airborne forces and was not intended for general issue.

Lewis Light Machine Gun

In 1939 the 1st Canadian Infantry Division crossed the ocean equipped with the Lewis Light Machine Gun instead of the modern Bren, and it was some time before issues of the Bren were completed. This was in spite of the fact that the Bren had been manufactured for the Canadian Army by the John Inglis Company of Toronto since 1938. By 1942, thanks largely to Canadian production, the provision of Bren guns had become much less of a problem. There were still some Lewis guns in the Corps, particularly in Corps troops, but 500 of these weapons had been disposed of to the War office on repayment, with more to follow as Brens became more available to Canadian formations.

By late 1942 the Bren had almost entirely replaced the Lewis light machine gun. At the beginning of 1944 only two Lewis guns were still held and these were museum exhibits at the Canadian Training School.

The Boys 55-calibre anti-tank gun shown here was effective against the earliest models of enemy tanks but was quickly rendered obsolescent. Library & Archives Canada

Bren Light Machine Gun

The Cameron Highlanders of Canada, who expressed so low an opinion of the Sten at Dieppe, reported on the same occasion that the Bren Light Machine Gun was a *"very accurate and efficient weapon"*. This appears to express the general opinion of the Bren gun held by personnel of the Canadian Army.

The Canadian Army Overseas was considerably better off than British formations, none of which had a complete establishment of Bren guns. By the end of January 1942, all of the Canadian Infantry Divisions were above establishment. For example, the 1st Canadian Infantry Division held 870 Brens against an establishment of 843.

However, by June 1942, establishments had been greatly increased and, for a time, supply was not adequate to meet increased demands. Mobilization entailed a very careful check of equipment by units affected. The result was that many guns were described as "beyond local repair" and were sent to the Canadian Ordnance Depot at Crookham. Most of these were repairable at Base Workshops and they were later placed back in stock. The mobilization plans also encouraged the British authorities to allot additional numbers of Brens to the Canadian Army Overseas. However, by April 1942 deficiencies still amounted to 19 percent of total war establishment. Finally, by 31 December 1943, deficiencies were virtually eliminated.

To increase the effectiveness of the Bren as an anti-aircraft weapon, a 100-round magazine was introduced in January 1943. A variety of special mounts were also developed to allow multiple Bren guns to be used against aircraft. With the authorization of the 20-mm anti-aircraft gun, the establishment of 100 round Bren magazine was deleted, although some 1,260 were still held by units in the United Kingdom in the summer of 1944.

The John Inglis Co. Ltd., of Toronto, produced over 186,000 Bren guns, in both .303 calibre for use by Commonwealth forces and in 7.92-mm for use by the Nationalist Chinese Army. Inglis produced both the Mark 1 and the Mark 2 versions. Two new models were produced in the United Kingdom, the Mark 3 and Mark 4 which were lightened versions of the Mark 1 and Mark 2 respectively and, like the light rifle, were developed primarily for use in the Far East. By war's end British production switched entirely to the Mark 3, while Canadian production remained on the Mark 2. The reason for Canadian production remaining on older

Soldiers of the Lake Superior Regiment (Motor) learning to fire the 2-inch mortar, November, 1942. Library & Archives Canada PA113367

models in the case of Rifle, Sten and Bren was that it depended mainly on British orders which were still for the older models.

Vickers Medium Machine Gun

The Vickers Medium Machine Gun (MMG) had originally played the role of a defensive weapon. Later emphasis on offensive training and mobility appeared to curtail the usefulness of this weapon. However, on-going experiments with a mounting device which allowed it to be fired from a carrier gave the Vickers a new purpose. The 1st Canadian Infantry Division took some of these new mounts to Sicily where it was used effectively from this mounting as well as in its more traditional role. There were numerous tributes to the effectiveness of the Vickers from all theatres. There was no shortage of Vickers guns or carriers during 1943. In August of that year 1st Canadian Infantry Division doubled its holding of MMGs.

As a result of the reorganization in January 1943, machine gun battalions disappeared from the Canadian Army Overseas and were replaced by divisional support battalions organized in three brigade support groups. Each of these groups consisted of a headquarters, a heavy mortar company, a 20-mm anti-aircraft company, and a medium machine gun company, the last-named armed with 12 Vickers guns. The effect of this reorganization was to reduce the total number of medium machine guns in an infantry division from 48 to 36. This reduction, plus an adequate and readily available supply from the British market, allowed for considerable reserves.

At the beginning of 1944 a further reorganisation of the divisional support battalions turned them once more into machine gun battalions. It was expected at first that this would restore the 48 guns per battalion, but in fact there was no change from the 36 guns formerly issued to the divisional support battalions. The new organization called for three companies each of 12 MMGs and one company of 4.2-inch mortars per battalion. In addition, independent machine-gun companies were formed for the infantry brigades of armoured divisions;

Members of the Queen's Own Rifles (3rd Canadian Infantry Division) firing a 3-inch mortar. Note the Mark III 'invasion' pattern helmet, England, April 1944 - Library & Archives Canada PA 197534

3-inch mortar training. Judging by the grins, poorly placed mortar and the fact that the soldiers are still wearing all of their kit, it can be safely assumed that this training photo was staged 'somewhere in England'. Library & Archives Canada e002343934

A 2-pounder gun on a Mark 2 carriage - identifiable by the pads at the end of the trails. The gun is probably a Mark 9A. The wheels have been removed for firing - one can be seen at the extreme right of the picture. Milart photo archives

these each had 12 MMGs and four 4.2-inch mortars. There were also eight MMGs in the motor battalion. Four of these weapons were allotted to the 1st Canadian Parachute Battalion.

2-inch Mortar

In late 1941 CMHQ reported that 2-inch mortars were 50 percent deficient of the establishment of 162 mortars. The report stated "*it appears that the deficiency in this field will be remedied only by the Canadian production now getting underway.*" This shortage continued into early 1942 when the 1st and 2nd Canadian Infantry Divisions each held 126 mortars while the 3rd Canadian Infantry Division held a mere 41. With the addition of the 4th Canadian Armoured Division, and the increase, through reorganization, of the number of units which used this weapon, the total requirement by 30 June 1943 was 1,348. The deficiency in the supply of 2-inch mortars was finally made up in March 1944. This deficiency was partly caused by the adoption of the 2-inch mortar in place of the smoke projector on carriers when it was discovered that the smoke projector was less efficient than a mortar mounted on the same vehicle.

The 2-inch mortar was found useful in a number of ways. The smoke bomb was employed to indicate targets to tanks, in addition to its usual job of supplying cover for the infantry. The high explosive bomb was also found to be effective and was carried in quantities almost as great as the smoke bomb.

3-inch Mortar

In late 1941 CMHQ reported that the issue of 3-inch mortars to the Canadian Army in the United Kingdom was almost complete. This weapon had been received in limited numbers during the summer of 1941, including some from Canadian sources. Experience at Dieppe directed attention to the fact that the British 3-inch mortar was seriously outranged by its German counterpart and attempts were made to increase its range. Later reports from Sicily indicated that, in spite of this disadvantage, it was a very valuable and extremely destructive weapon.

Reports from all fronts recognized the value of this mortar. Canadian units were supplied

A small number of carriers were modified to carry the 2-pounder anti-tank gun as a self-propelled platform. The Carrier, 2-Pounder, Attack, was limited to training roles in Canada and were never used operationally. MilArt photo archives. Library & Archives Canada e002505513

with a strengthened mortar (Mark 4) and a new base plate. There was no difficulty regarding supplies of this weapon or of carriers to transport it. Experiments were carried out in Canada with a lightened bipod and base plate as well as with a tapered barrel.

4.2-inch Heavy Mortar

The heavy mortar used by the Canadian Army Overseas was the British-produced 4.2-inch weapon. However, an alternative weapon, called the "Finnish" or 120-mm mortar, was also under development in Canada.

In December 1942, a CMHQ Equipment Policy Letter detailed the issue of heavy mortars, of either 4.2-inch or 120-mm calibre, to the Infantry Brigade Support Companies (later Support Groups). Each was allotted 12 of these mortars, as and when supplies became available. By the end of December 1942, five 4.2-inch mortars were available, but were not issued, as the Support Companies had not yet been set up. With the reorganization of the Army in January 1943, these companies came into existence and created a demand for 132 heavy mortars (36 per infantry division and 12 per armoured division).

The authorized allotment for Infantry Brigade Support Groups, published in March 1943, reduced the scale from 12 to eight mortars per group for a total of 66 mortars. This resulted in a surplus of 58 weapons as of December 1943.

At the beginning of 1944, the problem of transport for the 4.2-inch mortar had not been entirely settled. The intention was to use the new T16 Carrier to tow a 10-cwt trailer carrying the mortar. These carriers were not yet available. In February, stowage trials were started at the Canadian Training School. Towards the end of that month, authority was granted to modify the T16 to stow the mortar directly on the carrier. This method proved successful and, by April, a sufficient number of suitably modified carriers were available. This method was in use in North-West Europe by the Canadian forces only. In Italy, the mortar was carried on a trailer towed by the Loyd Carrier.

A very muddy 6-pounder anti-tank gun near Gouy, France, on 30 August 1944, giving a ride to several infantrymen. The heavily overloaded carrier has at least seven men on board and two more seem happy with their uncomfortable perch on the gun. Universal carrier and 6-pdr anti-tank gun of the Royal Winnpeg Rifles, 7th Canadian Infantry Brigade, 3rd Canadian Infantry Division, Normandy 1944. Library & Archives Canada PA132421.

The 6-pounder anti-tank gun, modified for a squeezebore "Littlejohn" adapter, which was screwed onto the threads that can be seen at the end of the barrel. It was an attempt to increase the muzzle velocity of the gun, but the adapter had to be removed in order to fire high explosive ammunition, and it never entered service. Library & Archives Canada e002343933

The 4.2-inch mortar played an effective part in operations. The one criticism was that the base-plate tended to bury itself in soft ground. This was corrected by the end of June 1944 with the issue of a base-plate skirting, a wide saucer-shaped rim fitting over the base-plate. This had the slight disadvantage of adding to the weight of the equipment.

A M14, 15cwt, half-track and 17-pounder anti-tank gun of an infantry division anti-Tank regiment - possibly belonging to 1st Anti-tank Regiment, RCA, 1st Canadian Infantry Division, Italy 1944. Library & Archives Canada e002505367

At the end of 1943 it was expected that the 95-mm Infantry Howitzer, then being developed, would replace the 4.2-inch mortar in infantry support groups. However, early in January this idea was abandoned when 21st Army Group chose to retain the 4.2-inch mortars. Development of the 95-mm Infantry Howitzer was finally stopped in December 1944.

Projector, Infantry, Anti-Tank (PIAT)

The Projector, Infantry, Anti-Tank (PIAT) was described as a shoulder-controlled weapon, designed as a replacement for the Boys anti-tank rifle, and capable of being loaded and fired by a single user. It weighed about 14 kg (31 lbs), and measured slightly less than 1 metre (approx. three feet). Two straps allowed it to be carried on one man's shoulders. It fired a bomb designed to penetrate armour. The weapon was produced in quantity, and the first supplies became available in October 1942. The first Canadian allocation of the PIAT was received in December 1942.

During succeeding months, supplies of the PIAT were delivered to the Canadian Army Overseas from British sources. However, owing to a serious shortage of ammunition, the decision to replace the Boys anti-tank rifle with these new projectors was not taken until April 1943. Canadian reports from Sicily noted that the PIAT was very effective against enemy tanks, subject to the reservation that it was necessary *"for the PIAT crew not to fire until they can literally almost see the whites of the enemy's eyes."*

In Italy there was still a deficiency of these weapons at the beginning of the 1943, but by the end of April there were sufficient stocks in the theatre to cover establishments and supplies increased each month. In North-West Europe and the United Kingdom there was no deficiency. The ammunition supply had definitely improved, although until near the end of the year operational requirements somewhat restricted the amount available for training. As late as September 1944 training units were still limited to one bomb per man.

Reports from operations testified to the effectiveness of the PIAT against pill-boxes and in a house-breaking role. The adoption of a new graze fuse for the bomb considerably improved its performance against tanks. It was even found effective against Tiger tanks.

A 17-pounder anti-tank gun of the Royal Canadian Artillery in Normandy, June 1944. This was the Allies' heaviest anti-tank gun of the war and was effective against any German tank. From the angle of the artillery trailer, the crew are trying to back the gun tractor-trailer-gun combination out of the alley - a process that was much like pushing string. Library & Archives Canada PA169273.

Boys Anti-tank Rifle

As late as the summer of 1941 there were still serious deficiencies of certain items of infantry armament. This was particularly true in the case of the Boys Anti-tank Rifle (.55-inch) As of 31 August 1941, units of the Canadian Corps possessed only 563 of these weapons compared to an establishment of 1,318. Nor did this situation show any signs of early improvement. The only anti-tank rifles shown on issue in the Corps General Report up to June 1941 were twelve Boys rifles doled out to the 3rd Canadian Infantry Division. During August, General McNaughton gave some aid to the newly arrived formations from the slender stocks of 1st and 2nd Canadian Infantry Divisions; nine were transferred from these divisions to the 1st Canadian Army Tank Brigade, and 48 to the 3rd Canadian Infantry Division.

In the Dieppe operation, because of its heavy calibre and extreme accuracy, the Boys rifle was found very useful for dealing with snipers dug in behind light protection. It could also penetrate light armour at short range. Presumably for these reasons, some of these weapons were taken by the 1st Canadian Infantry Division to Sicily.

Thanks largely to increasing Canadian production at the John Inglis Company in Toronto, the deficiency of Boys anti-tank rifles was progressively reduced during 1942. In December 1942, the Boys .55-inch anti-tank rifle was withdrawn from all units equipped with 2-pounder anti-tank guns or awaiting issue of 20-mm guns. This change reduced establishments of Boys rifles by 1,322 weapons. In effect, a deficiency of 565 weapons had become a surplus of 1,193. The Boys was, however, to be retained in the fighting vehicles of armoured car and reconnaissance regiments as well as infantry and motor battalions *"and by units which will not be issued with 20 mm guns"*. The decision to withdraw the Boys from the units mentioned (in December 1942) had been undertaken in anticipation of soon receiving large numbers of 2-pounders, 6-pounders and 20-mm guns. When the hoped-for supplies of 6-pounders and 20-mm guns did not appear the Boys was re-issued. The re-issued Boys were to be replaced with the PIAT on a *pro rata* basis as supplies made it practicable. The surplus stock of the Boys anti-tank rifle was transferred to British ordnance depots.

A 25-pounder Mark I (18/25-pounder) on a Mark IVP carriage. This was the 18-pounder conversion to the 25-pounder and was used by artillery regiments of the 1st Canadian Infantry Division in 1940. This clearly shows the positions in action of the six-man detachment (the three standing men are not part of the detachment). Pictured is No. 3 gun, 'E' Troop, 'C' Battery, RCHA, 1st Field Regiment, RCHA, 1st Canadian Infantry Division, Barham, Sussex, 10 April, 1942. Library & Archives Canada PA131091

2-pounder Anti-tank Gun

The situation with respect to 2-pounder anti-tank guns had been very difficult in the early days of the war. When approved in 1936, the 2-pounder anti-tank gun was among the best in the world. By 1940 it was effective against the German tanks encountered, but even then was obsolescent. When more than 500 2-pounders were abandoned in France in 1940, it was kept in production on the basis that a gun available now was more desirable than a better gun tomorrow. It had a maximum range of 8,000 yards (7,309 metres) and could penetrate 42 millimetres of armour at 1,000 yards (913 metres). The 2-pounder was manufactured in both Britain and Canada.

There had been plenty of carriages available but a great lack of barrels. These were going to the arming of tanks, which took priority over the arming of anti-tank units. This situation was reflected in 2nd Canadian Infantry Division. The Division's anti-tank unit (2nd Canadian Anti-tank Regiment) received its first three 2-pounder guns only during the week ending 18 January 1941. Nine more were issued during the week ending 15 February 1941, and twelve were transferred from 1st Canadian Infantry Division. By March 1941, 2nd Canadian Anti-tank Regiment still possessed only 36 guns, and 1st Canadian Anti-tank Regiment was still short the 12 guns which it gave up. On the other hand, a relative improvement in this field may be apparent in the fact that the 7th Canadian Anti-tank Regiment, organized as a Corps Troops unit, had 24 guns - but 12 of these were transferred to the 3rd Canadian Anti-tank Regiment when it arrived in England. It was clear that the situation was far from satisfactory.

A 25-pounder Mark II on a Mark 1 carriage. The man has a rammer under his arm. The ammunition was separately loaded (the shell was placed in the breech and rammed home, then the cartridge case containing the propellant was loaded and the breech was closed). The artillery limber, with its ammunition trays, is in the background. Library & Archives Canada e002107897

In February 1942, the 2-pounder had become available in somewhat greater numbers, and each overseas division had 36 of these guns. In addition, 3rd Canadian Anti-tank Regiment also held five 2-pounder guns mounted on carriers for experimental purposes. Nonetheless, a deficiency existed due to a change in establishment which saw the divisional allotment increased from 48 guns to 64. This continued deficiency led to the policy of replacing all 2-pounder guns in anti-tank regiments with the 6-pounder anti-tank gun. For the 2-pounders withdrawn from anti-tank regiments other uses were found. CMHQ in September 1942, announced that 2-pounders would be issued to infantry battalions, motor battalions and reconnaissance regiments, on a basis of six guns to each infantry or motor battalion and twelve to each reconnaissance regiment. With the increased availability of 6-pounder equipment, however, the 2-pounder disappeared even from the establishments of infantry anti-tank platoons. CMHQ, on 9 June 1943, advised that 2-pounders issued to the above mentioned units would be withdrawn to reserve and replaced by 6-pounders on the same scale (subsequently increased from six to twelve guns in the case of motor battalions). The 2-pounders were withdrawn as the 6-pounders were issued.

In 1942, experiments were conducted with a high-velocity 2-pounder gun - the "David". This weapon used a very powerful charge to fire a "composite-rigid" shot, the outer shell of which disintegrated on impact, penetration being made by a small tungsten carbide bullet. But its adoption by the Canadian Army depended on similar action by the British, and it was considered unlikely that they would find any use for it. In Italy, some 2-pounder taper-bore anti-tank guns ("Littlejohns") were issued to Canadian units for river crossings and mountain fighting. They were used for the first time in October. Ultimately, the 2-pounder was withdrawn from service in favour of the newer and more powerful 6-pounder.

US-built M-10, 17-pounder self-propelled gun of the 63rd Anti-Tank Regiment, R.A., carrying troopers of the 1st Canadian Parachute Battalion (British 3rd Parachute Brigade, 6th Airborne Division), near Hanover, Germany, 7 April 1945. Library & Archives Canada PA191134

6-Pounder Anti-tank Gun

The 6-pounder anti-tank gun was designed in 1938, but did not enter into immediate production. Although the need was clear in 1940, the 2-pounder was kept in production, and the first 6-pounders did not leave the factory until November 1941. The gun could penetrate 74 millimetres of armour at 1,000 yards (913 metres) and, later in the war, improved ammunition was effective even against Tiger tanks. The 6-pounder was manufactured in Britain and in Canada. It was mounted in tanks and also on a towed carriage.

The intention of re-arming anti-tank regiments with the "*new and powerful*" 6-pounder gun, was duly carried out during the summer and autumn of 1942. As early as January 1942 it had been decided to replace all 2-pounders in the Canadian Corps with 6-pounders and a new war establishment, effective 15 March 1942, contained the provision that the armament of an anti-tank regiment would be sixty-four 6-pounders. By mid-July 1943, the 2nd and 3rd Canadian Infantry Divisions and the 5th Canadian Armoured Division were complete in 6-pounders.

By 1944, the 17-pounder and M-10 self-propelled anti-tank guns had largely replaced the six-pounder in the artillery anti-tank regiments. However, those assigned to an infantry division retained one troop of four 6-pounders in each battery (a total of 16 guns per regiment). Only the 2nd Canadian Infantry Division used exactly this establishment. The 3rd Canadian Infantry Division held a higher proportion of 6-pounders in place of 17-pounders for the assault in June 1944, and they retained these for the rest of the war. In Italy, the anti-tank regiment in the 1st Canadian Infantry Division was organised into two 6-pounder troops and one 17-pounder troop per battery. Also, an experiment in Italy temporarily replaced the anti-tank platoon of an infantry battalion with an anti-tank company that used No.75 grenades, PIATs,

A Sexton self-propelled field gun. Designed and manufactured in Canada, the Sexton successfully mounted a 25-pounder gun on a Ram tank chassis. 2,150 were produced and it became the standard SP gun in the British Commonwealth armies.Library & Archives Canada e002505354

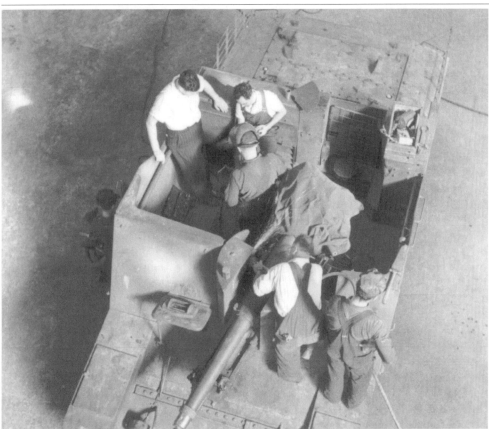

Workmen at Montreal Locomotive Works finishing a Sexton 25-pounder self-propelled gun. The gun barrel is threaded for a muzzle brake that has not yet been fitted. This vehicle was a Canadian development that met with great success in both the British and Canadian armies. Library & Archives Canada

A rare photograph of the Sexton GPO (Gun Position Officer). A dozen Sexton 25-pounder self-propelled guns were modified, in England, into GPO vehicles by removing the gun, closing the gun port and installing a gun-laying table and additional radios. 23rd Field Regiment (SP), RCA, 4th Canadian Armoured Division, Netherlands, 8 June 1945. GPO Vehicle 'Darling', T-204815; Courtesy Dana Nield

and 2-pounder anti-tank guns. This was a special arrangement for the type of operations which were being undertaken.

4th Canadian Anti-tank Regiment of the 5th Canadian Armoured Division in the same way continued to use some 6-pounders. There were no deficiencies of this weapon in North-West Europe. In Italy, a slight deficiency was covered by the use of the US 57-mm M1 anti-tank gun. This was the American counterpart of the 6-pounder. The deficiency was made up by the end of the year.

The 6-pounder was towed by the new T16 Carrier in reconnaissance regiments, motor battalions, and infantry battalions in North-West Europe. The Portee (carrying the gun in the back of a truck) used in Italy was not found satisfactory. By October 1943, this was being replaced by half-track vehicles. In anti-tank regiments the gun was towed by the field artillery tractor.

In operations the gun was found particularly effective against buildings. A report from Italy states that it was more effective than the 17-pounder in this role. In North-West Europe it was found very effective against tanks when using the new discarding sabot ammunition.

17-Pounder Anti-tank Gun

The 17-pounder anti-tank gun was designed in 1941, and entered British service in late 1942. Although a very large gun on its towed carriage, it was justified because, using discarding sabot ammunition (penetrating 231 millimetres at 1,000 yards (913 metres)) it out-matched any German tank in the war. It was also used mounted in a self-propelled chassis, and as the main gun in the "Firefly" version of the Sherman tank.

Once the 17-pounder became available, anti-tank regiments were armed with a mixture of 6-pounder and 17-pounder guns. War Office policy, as notified on 6 April 1943, was that the proportion of guns within regiments should be *"one-third 17-pounder to two-thirds 6-*

A 105-mm M7 SP gun (Priest). The Priest was used by the field regiments of Royal Canadian Artillery for the beach assault in Normandy, and also by the SP regiment in Italy. After the assault, it was replaced by towed 25-pounders or by Sextons. The regiment in Italy converted to Sexton when they moved to France in 1945. Pictured is No. 4 gun, 'E' Troop, 78th Field Battery, RCA, 13th Field Regiment, RCA, 3rd Canadian Infantry Division in Normandy, July 1944. Library & Archives Canada PA114577

pounder". In the following July it was announced that British anti-tank organization in infantry and 'mixed' divisions would be based upon a regiment of four batteries, each of two troops of 6-pounders and one troop of 17-pounders. For a time, Canadian units were equipped upon a somewhat different basis, the anti-tank regiments of 2nd and 3rd Canadian Infantry Divisions each having three 6-pounder batteries and one 17-pounder battery. 1st Canadian Anti-tank Regiment (1st Canadian Infantry Division), which was employed with the Eighth Army in Italy, was organized on the British basis of composite batteries. Subsequently 2nd Canadian Anti-tank Regiment was likewise reorganized. Canadian units were complete to establishment in 17-pounders by 30 June 1943.

Beginning early in 1944 Canadian anti-tank regiments were equipped with the towed 17-pounder anti-tank gun on the basis of two troops per battery in infantry divisional anti-tank regiments, and two batteries per regiment in armoured divisions and corps troops. This establishment was in fact subject to modification. 3rd Canadian Infantry Division had never received any towed 17-pounders since their assault landing in June, preferring to retain the 3-inch M10 self-propelled guns issued for that occasion. There were no problems with supply as the weapons were obtained from British sources as required.

The question of a tow-vehicle (termed 'tower' in contemporary documents and used here) for the gun received considerable attention. Within 21st Army Group the policy was adopted of fitting all tanks with 17-pounder towing hooks for emergency use. In March 1944 it was intended to use entirely wheeled vehicles as standard towers, with the American 2fi-ton 6x6 lorry being considered. Although these were used by the British in Italy, there were insufficient available for use by Canada and various other vehicles were used. 6th Canadian Anti-tank Regiment of 2nd Canadian Corps used Ram tanks modified for this purpose. They were to change to Crusader towers but this never occured. In Italy, the anti-tank regiments of the 5th Canadian Armoured Division and 1st Canadian Corps used modified Crusader tanks. Other units used the field artillery tractor or half-tracks.

The 17-pounder anti-tank gun proved very effective in operations. But there was a defi-

M7 'Priest' 105mm Howitzer Self-Propelled Gun. The vehicle pictured belongs to the 98th (Surrey & Sussex Yeomanry) Army Field Regiment (SP), RA, who were 8th Army Troops in Italy (where this photo was taken) - and is identified as No. 1 Gun, 'C' Troop, 'Q' Battery. The 98th supported the 1st Canadian Infantry Division's attack across the Moro River on 5/6 December, 1943. Library & Archives Canada PA177534

nite preference for self-propelled guns. The towed gun was hard to move and a tracked tower was often necessary. The Ram tower was reported to be the best available, but was noisy and bulky.

M10 Self-propelled Gun

The US Gun Motor Carriage M10 was a self-propelled anti-tank gun that married a 3-inch anti-aircraft gun to a Sherman tank chassis. The British bought over 1,600 of these vehicles and modified 1,000 of them to accept 17-pounder guns. The Americans called it a "tank destroyer" and had problems with their doctrine for its use, but in the British and Canadian forces it was simply considered to be a self-propelled anti-tank gun.

These self-propelled anti-tank guns were issued to the Canadian Army as an interim measure pending the availability of self-propelled 17-pounders. They were issued to two batteries per regiment in armoured divisions and corps troops. In addition, one troop per battery of 3rd Canadian Anti-tank Regiment of 3rd Canadian Infantry Division received them for the assault landing in June. By the end of February 1943, the Canadian forces in the United Kingdom were fully equipped on this basis. The Canadian forces in Italy received their quota in April.

The anti-tank regiments of the armoured divisions and Corps Troops consisted of two 17-pounder batteries and two 3-inch M10 SP batteries. In the case of 3rd Canadian Anti-tank Regiment, this formidable self-propelled weapon was to be exchanged for tractor-drawn 6- and 17-pounders when its D-day assault task had been performed. By 31 December 1943 this unit held the 16 SP guns authorized for it.

In June 1943 twenty-four of the M10/17-pounders were issued to 5th Canadian Anti-tank Regiment of 4th Canadian Armoured Division in place of the 3-inch M10 equipment which had been held as an interim measure. At the end of June three more were issued to Canadian reinforcement units for training.

The Archer 17-pounder self-propelled anti-tank gun. This mounted the gun facing rearwards on a Valentine tank chassis. Space was so limited that the driver had to get out of his seat before firing to prevent being decapitated by the recoil. Although they appear awkward, they were small, mobile, and easily concealed, and were very successful in operations. Library & Archives Canada PA116841

Self-propelled 17-pounder, Valentine Mark I, Archer, anti-tank gun of the 7th Anti-Tank Regiment, RCA, 1st Canadian Corps Troops, Cesena, Italy 20 October 1944. Library & Archives Canada PA173754

It was possible in April to allow eight M10 self-propelled equipments in each corps and armoured divisional anti-tank regiment in First Canadian Army to be run to death and replaced. Apart from the gradual replacement by 17-pounder self-propelled guns there was no change in the holdings of this weapon. 3rd Canadian Anti-tank Regiment did not replace their self-propelled guns with towed 17-pounders after the assault as originally intended; in August, in place of 3-inch equipment, they received 17-pounder M10 equipment.

A 6-inch 26-cwt howitzer on a Mark 1P carriage. The gun dated from 1915, and was obsolete in 1939, but was still taken to the UK by 1st Medium Regiment, RCA, when war broke out. It was replaced by the 5.5-inch gun, but remained in service for training purposes for much of the war. Library & Archives Canada e002344116

All reports from the field spoke highly of the self-propelled anti-tank guns. They proved themselves very effective against enemy tanks. But the great value of the self-propelled gun was its ability to get forward quickly to give close support to the infantry. On at least one occasion the M10's proved their ability to travel long distances at considerable speed.

25-pounder Field Gun

The 25-pounder field gun was designed in the late 1930s to replace the 18-pounder of First World War vintage. It fired a 25-pound shell (high explosive, smoke, or other variants) to a maximum range of 13,400 yards (12,244 metres). In 1939, Canada built a plant in Sorel, Quebec to manufacture the gun, with the first weapon being delivered in July 1941. The 25-pounder was the standard field gun in the Commonwealth, with more than 17,700 being produced, 3,781 of them in Canada.

The towed 25-pounder equipped three 24-gun field regiments in each infantry division, one regiment in each armoured division, and others in the Corps Troops. Although early supplies were a problem, by February 1942, all field regiments possessed their full complement. This included 7th Field Regiment that had only arrived in the UK in November 1941. By 30 June 1943, holdings of towed 25-pounders (including reserves) were virtually complete and, by November 1944, production of the gun was actually surplus to requirements.

Sexton Self-propelled Gun

There was a special development in the equipment of field artillery regiments when the decision was made to provide one of the two field regiments in each armoured division with self-propelled 25-pounders. These weapons were the Sexton - 25-pounder field guns mounted on a Ram tank chassis - and were produced in Canada. Initial supply was slow and 8th Canadian Field Regiment (5th Canadian Armoured Division) was reported, on 1 November 1943, as holding only 18 Sextons against an establishment of 24. These were left behind, along with all other heavy equipment, when the division embarked for the Mediterranean theatre.

The 23rd Canadian Field Regiment of the 4th Canadian Armoured Division had already

At least one regiment's worth of 5.5-inch medium guns. There were problems with barrel wear, but the gun was reasonably well liked, firing a 100-pound (45.5 kg) shell to 16,200 yards (14,800 metres). Library & Archives Canada

obtained its complement of Sextons in April 1943, while still in Canada. The Canadian forces in the United Kingdom actually held 32 of these weapons at the beginning of the 1944. The rest of the Canadian production was consigned to the British who would provide for necessary replacements in the Canadian Army. Canadian supplies were sufficient for British needs and, in August, they were willing to have the production of Sextons cut to allow the Canadian factory (Montreal Locomotive Works) to turn its attention to producing the Ram Observation Post (OP) vehicles. These latter were Ram tanks with the guns replaced by a false barrel and the addition of an extra radio and a plotting board.

A dozen Sextons were built as command vehicles for gun position officers. These Sexton GPOs were provided by conversion of standard Sextons in England - the production of Sextons in Canada was not affected. In Italy, the 5th Canadian Armoured Division received, from the British, 24 Sextons in April. On the formation of 12th Canadian Infantry Brigade these were replaced by towed 25-pounders.

105-mm Self-propelled Gun M7 (Priest)

Another development was the introduction of an American self-propelled field piece - the 105-mm M7 gun on a Sherman tank chassis, called the Priest. Like the 25-pounder, the 105-mm was a gun-howitzer, but larger, and used seven different propellent charges, the British piece using only four. There was no firm war establishment for these, but they were issued to five field regiments (including the three regiments of 3rd Canadian Infantry Division) for their special beach assault role. Provision was to be from British sources. On 1 December 1943, the Canadian Army Overseas held 60 Priests.

There had been some initial delay in obtaining these. Some, originally intended for the Canadian Army, were diverted to the Mediterranean. But by the end of February 1944, the self-propelled artillery regiments were completely equipped. Self-propelled guns were used for the assault landing as they alone could fire from the landing craft while on the way in to shore and still be easily moved when they landed muzzle first. These guns were to be replaced after the landing with tractor-drawn 25-pounder guns. The replacement duly took place in August for the three regiments of 3rd Canadian Infantry Division. At the end of August, 19th Canadian

A 6-inch 26-cwt howitzer on a Mark 1P carriage. Library & Archives Canada e002505359

Field Regiment exchanged its Priests for 25-pounder Sextons. In April 8th Canadian Field Regiment (SP) of 5th Canadian Armoured Division received 24 Priests and these were still in use at the end of the year. The Priest was most popular in Italy.

In August 1944, 72 of the replaced Priests in North-West Europe were converted into armoured personnel carriers by removing the guns and, under the name 'Kangaroo', rendered valuable service. Without their guns they were humourously termed 'unfrocked Priests'. The guns were later replaced and the equipment returned to the Americans. The need for a Kangaroo remained, however, and a large number of Ram tanks were then converted to this role.

Centaur 95-mm

There was, for a very short period of time, a Royal Canadian Artillery Battery issued with the Centaur self-propelled gun. The 1st Canadian Centaur Battery was formed on 6 August 1944 and disbanded on 30 August 1944. Manned by Canadian reinforcements and British Royal Artillery personnel, it consisted of three troops and a Headquarters. The establishment for each troop was a Sherman OP (Observation Post tank) and three Centaur Mark 4. The Battery saw action in mid-August and was disbanded at the end of the same month.

Medium Artillery - 4.5-inch and 5.5-inch Guns

The 1st Medium Regiment, RCA, arrived in Britain in 1939 fully equipped with obsolete 6-inch howitzers that they brought from Canada. However, medium guns were in particularly short supply and no other medium regiments had been brought to England. In light of General McNaughton's views on the utility of concentrations of medium and heavy guns (well known to all students of his writings), and the fact that long-term plans called for five medium regiments in the Canadian Corps, the situation was absurd and every effort was made to improve it.

To replace the 6-inch howitzer and the 60-pounder gun, Britain designed the 4.5-inch and 5.5-inch medium guns in 1938-9, but production was slow and neither was quick into service. Apart from the barrel, the guns and carriages were identical. The 4.5-inch gun fired a 55-pound

Bofors 40-mm anti-aircraft guns preparing for action. There were two gunlayers, one laid the gun in elevation and one for traverse. Each had his own sight. Library & Archives Canada

(25 kg) shell to 20,500 yards (18,274 metres) while the 5.5-inch gun fired a 100-pound (45.5 kg) shell to 16,200 yards (14,800 metres). Although most Canadian medium regiments eventually used the 5.5-inch, many medium gunners preferred the 4.5-inch gun, especially for mobile operations, because it had a longer range and was more accurate.

In March 1941, following General McNaughton's representation, the possibility of manufacturing the new 5.5 inch gun-howitzer in Canada had been explored by the Department of Munitions and Supply. The Department decided that manufacture of barrels in Canada was impracticable, even though Canadian National Railways Munitions Ltd., had been at work on a British order for 300 5.5-inch carriages. In July, a complete review noted that the United States had proposed to supply Britain with 300 4.7-inch US barrels that had been relined to 4.5-inch and fitted to the British 5.5-inch carriage. Under these conditions, any provision of medium guns for the Canadian Army, from American sources, depended on obtaining a release from the British War Office, which agreed to fill Canadian overseas requirements for the 5.5-inch gun and carriage from UK sources. In fact, four 5.5-inch guns were released immediately to the Canadian Corps for user trials. The 7th Medium Battery, less one troop, of 5th Medium Regiment, RCA, was organized in England to conduct the trial. The intent was to organize the remainder of this regiment as guns became available. This was delayed when, by the summer of 1941, the first four guns had not yet been received. However, the supply situation soon improved and the regiment had their complete establishment of sixteen guns by mid-October. At the time it was also hoped to replace 1st Medium Regiment's 6-inch howitzers with 5.5-inch guns. The prospect of obtaining additional guns led to the creation of the 3rd and 4th Medium Regiments, in Canada, with a view to their despatch to Britain in the spring of 1942.

By the end of 1943 there were six Canadian Medium Regiments overseas, of which three were in Italy. Deficiencies of 5.5-inch gun-howitzers amounted to 16 guns as at 30 September 1943. No figures were, at the time, available for the three regiments in Italy, but by 31 December 1943, those units remaining in the United Kingdom possessed their full allotment of 16 guns each.

The three regiments in the United Kingdom at the beginning of 1944 were fully equipped with 5.5-inch guns during most of the year. Although the expenditure of barrels in First

Bofors 40-mm anti-aircraft guns being demonstrated to senior officers, Colchester, 9 July 1941. Library & Archives Canada PA167590

Canadian Army caused trouble at one period, there was no deficiency of these weapons. At the end of July it had been decided to replace them in one regiment with 4.5-inch guns due to their greater range. By the end of December 1944 this replacement was complete.

In Italy there was considerable delay in equipping the three Canadian medium regiments. Only 1st Canadian Medium Regiment was equipped by the end of January 1944. In February the other two regiments began to receive 4.5-inch guns in lieu of 5.5-inch. By the end of the month the deficiency had been made up by further issues of 4.5-inch guns. In August, a further issue of 5.5-inch guns allowed two regiments to be equipped with this weapon, leaving one with the 4.5-inch guns.

20mm Anti-aircraft Artillery

The position of the 20-mm gun within the Canadian Army changed considerably during 1944. To meet immediate requirements, M1 Hispano guns on American M2A1 mounts were issued. These were later to be replaced by Canadian Inglis mounts as they became available, and Hispano M2 guns were ordered from the United States for use on these Inglis mounts. Finally, Inglis 20-mm guns would be used on the Inglis mounts. In anticipation 2,600 single mounts, and 1,000 quadruple mounts were ordered.

Early in 1944, there were 1,950 M2A1 mounts available, and an equivalent number of Hispano Mark 1 guns were being obtained from the British in place of the American M1. These needed modifications, which presented some difficulties, but issues began in January 1944. By February, the Inglis quadruple mounts were entering production. A total of 6,600 Hispano M2 guns had been ordered from the United States for use on these mounts and 3,800 had been delivered to Canada by February. The Inglis gun, which would use Hispano type ammunition, was still in the development stage. The guns on the single mounts were allotted on a generous scale to a wide variety of units, the total establishment in First Canadian Army being 1,065. The quadruple mounts were intended for light anti-aircraft regiments.

Early in February 1944, this policy had to be extensively revised when the Commander-in-Chief, 21st Army Group, decided not to use 20-mm guns, except in one troop per battery of the divisional light anti-aircraft regiments, and on anti-aircraft tanks and armoured cars. This

Triple-mount 20-mm Polsten anti-aircraft gun. This weapon was used as a self-propelled model as well as on a variety of trailers (see cover for alternate trailer). Library & Archives Canada

virtually eliminated all requirements for quadruple mountings and reduced the need for single mountings from over 1,000 to only 72.

Canadian policy was to conform to the British equipment standards in order to simplify maintenance and ammunition supply, particularly since First Canadian Army was a composite force. The 20-mm guns had been an exception to this because the Canadian Army preferred the Hispano type to the Oerlikon type adopted by the British. But in light of the greatly reduced requirement, this was no longer practical, and it was decided to adopt the British Polsten on the Universal mounting (which could accommodate all three types) for First Canadian Army. Canada was advised to cancel as much as possible of the Hispano M2 and Inglis single mount orders, and not to start production of the Inglis gun. The Inglis quadruple mountings were still to be produced, but for use with Polsten guns. The delivery of Hispano M2 guns was stopped at 3,800 and it was found possible to cancel the whole order for Inglis single mounts.

The new establishment for 20-mm guns called for only one troop of eight guns per battery in divisional light anti-aircraft regiments. By April 1944, these troops were completely equipped with Polstens on Universal mountings, though training in Canadian reinforcement units was still carried out mainly with the Hispano Mark 1. The Canadian forces in Italy were equipped throughout this period with Oerlikon 20-mm guns on the Hazard-Baird mounting. The 20-mm gun was also mounted on tanks for use in armoured units. By June, the first deliveries of the quadruple mounts arrived in the United Kingdom, but there was no immediate use for them. Some were self-propelled on a 3-ton chassis, while others were towed on a travelling platform. One was tried against flying bombs in England, but otherwise no use was found for them and, in August, it was recommended that the rest of the order be cancelled due to the vast improvement in the strategic situation in the air since these equipments were ordered. An outright cancellation was found to be uneconomical and the contract was in fact completed. Altogether 441 mounts were shipped to the United Kingdom from Canada.

Ford 3-ton truck with 20mm Polsten quad anti-aircraft gun. Compare this configuration to the prototype Inglis model on page 61. Library & Archives Canada e002505370

In the final analysis, very little use was found for the 20-mm gun in an anti-aircraft role. One great difficulty was the lack of self-destroying ammunition, which caused considerable danger to our own troops in the neighbourhood. As early as October 1943, Canadian units in Italy were considering using it in a ground role, but no suitable mounting had then been developed, although local experiments continued. Similarly in Northwest Europe, the 20-mm gun was soon believed to be unnecessary. Anti-aircraft tanks were no longer being supplied, although 10th Canadian Armoured Regiment had found their guns useful in a ground role. By August 1944, the 20-mm troops in the light anti-aircraft regiments were being disbanded. There was renewed interest early in November 1944, when the activities of the German Me-262 jet-propelled aircraft created a demand for anti-aircraft weapons. Four of the once-unwanted, trailer-borne, quadruple mounts were requested by First Canadian Army. Early in December two self-propelled and two trailer-borne mountings were sent for trials. Similar issues were made to the British in Italy and India. Finally, in December it was suggested that the M2A1 mountings be disposed of as salvage.

Though it would appear that a great deal of effort and expense was incurred in developing a weapon which was scarcely used, this was not quite the true picture. The Allies gained air superiority and there was a definite improvement in the strategic situation while the weapon was in production, resulting in a decrease in requirements. When not needed in an anti-aircraft role it was used as a heavy machine gun in a ground role. Large numbers of 20-mm guns were used on ships, where the air attack threat was quite different.

40-mm Light Anti-aircraft Artillery

The Bofors 40-mm anti-aircraft gun was developed in Sweden in the 1920s, and bought by Britain in 1937, with the purchase including a licence to manufacture the weapon domestically. It fired a 2-pound (0.9 kg) shell to an effective ceiling of 5,000 feet (1,500 metres) with a rate of fire of 120 rounds per minute. Canada manufactured large numbers of Bofors guns, on many different mountings, starting in 1940.

Lorry, 3-Ton, 40-mm, Self-propelled - used by light anti-aircraft Regiments, RCA. A Bofors 40-mm anti-aircraft gun on a Ford S-60B chassis. Note the spare barrel mounted on the fender. Milart photo archives

Inglis quad-mount 20-mm anti-aircraft gun. Although a large quantity were ordered the contract was cancelled before deliveries in quantity could start. Photo Globe & Mail Collection, City of Toronto Archives, courtesy Blake Stevens.

The Canadian anti-aircraft and anti-tank group at Colchester possessed, in September 1941, a total of twenty 40-mm Bofors. This was a respectable provision when it is considered that the majority of the light anti-aircraft batteries formerly there were performing operational roles under Air Defence Great Britain (ADGB) Command, which provided the guns for these purposes. An arrangement had been made by which, on the "Stand To" signal being received, Headquarters Canadian Anti-aircraft and Anti-tank Group would provide for service with the

A 3.7-inch heavy anti-aircraft gun on a Mark 2 static mount. The gun is employed in the coastal defence of Britain. Library & Archives Canada e002343946

Canadian Corps, one light anti-aircraft battery, and a second battery, less one troop, armed with these 20 guns.

Apart from the needs of the Middle East, it was understood that part of the production of Bofors had been diverted for mounting on vessels of the Merchant Navy.

The total holdings of Canadian anti-aircraft units by 31 January 1942, had increased to 58 guns. As a result of this situation the 2nd Canadian Light Anti-aircraft Regiment was able

A 3.7-inch heavy anti-aircraft gun, Mark 3A. This was the standard heavy AA gun in the British and Canadian forces, and was manufactured in Britain and in Canada. The gun served in 2nd Heavy AA Regiment, RCA, in Europe, and was also used for the air defence of ports and other critical locations in Canada. Photo courtesy DND, Directorate of History & Heritage.

Workmen fire a 3.7 gun at the Valcartier test ranges, February 1943 at Valcartier, Quebec. National Film Board collection. Library & Archives Canada e000761174

A finished 40-mm Bofors anti-aircraft gun, probably at the Otis-Fensom Elevator Company factory. Canada shipped the Bofors to the US, Britain, Middle East, and naval establishments around the world. Library & Archives Canada

to take its place in the Order of Battle in the 1st Canadian Infantry Division of the Canadian Corps.

The shortage of light anti-aircraft artillery weapons mentioned earlier persisted throughout 1942. In October, the six light anti-aircraft regiments overseas were only 50 percent complete in Bofors 40-mm guns. Canada was at that time producing about 200 of these weapons per month, but shipping allotments were such that the guns were not being despatched to the Canadian Army Overseas. Strong requests for more guns were made to the London Munitions Assignment Board. In consequence, supplies became available in increasing numbers and deficiencies were made up by March 1943, despite the fact that establishments had increased until the total for the Canadian Army Overseas was 519 guns. By summer 1943, all nine reg-

A Tillings-Stephens production version of the Land Mattress. The rocket projector was unique to the Canadian Army and fired 30 rockets in eight seconds, producing a massive impact on the target. The rockets were assembled largely from surplus and scrapped parts. Library & Archives Canada

iments overseas had their full complement of 54 guns each. By October 1943, 61 percent of 40-mm equipment was of British origin, the remainder being Canadian.

Very little opportunity was found for the Bofors in an anti-aircraft role in operations due to the weakness of the enemy air force, and reports speak mostly of its services in a ground role. In this role it was used both in Italy and in North-West Europe for ordinary ground shooting, using tracer to mark lanes for night attack and using armour-piercing and high explosive to blow in buildings.

Supplies of the Bofors 40-mm gun were adequate throughout 1944. For the purpose of the assault landing, 4th Canadian LAA Regiment of the 3rd Canadian Infantry Division was completely equipped with self-propelled anti-aircraft guns holding 54 by the end of February. The conversion of one troop per battery for the other regiments proceeded steadily. 3rd Canadian LAA Regiment of 2nd Canadian Infantry Division was complete by the end of March as was 6th Canadian LAA Regiment of 2nd Canadian Corps. 8th Canadian LAA Regiment of 4th Canadian Armoured Division were not fully equipped with self-propelled anti-aircraft guns until July, but they held a surplus of tractor-drawn guns in lieu.

In Italy, 1st and 5th Canadian LAA Regiments received the self-propelled guns in April and May, respectively, and 2nd Canadian LAA Regiment was equipped with these soon afterwards. During 1944 there was a tendency to reduce the establishment for light anti-aircraft weapons and, in March 1944, 7th and 11th LAA Regiments were disbanded. In Italy this reduction made it possible to convert 1st Canadian LAA Regiment to an infantry battalion as part of 12th Canadian Infantry Brigade.

The general trend towards self-propelled anti-aircraft artillery pieces was highlighted in the announcement, in December 1943, that LAA Regiments would convert one troop per battery to 40-mm self-propelled equipments. The Canadian Army in the United Kingdom already held forty-two 40-mm SP guns mounted on Morris 30-cwt 4x2 trucks. These would be replaced by Ford 3-ton trucks during the summer of 1943. Some guns mounted on the Morris

Lifebouy flame-thower. The name was derived from the unique shape of the fuel container. The Lifebouy was used with impressive results during the fighting in the Netherlands. MilArt photo archive.

chassis were still held in First Canadian Army, but this was not due to a lack of the Canadian vehicles. In September it was considered that, in view of the diminishing requirements, further Canadian vehicles would not be needed. Of those held in Canada, 52 were diverted to Italy

3.7-inch Heavy Anti-aircraft Artillery.

The 3.7-inch heavy anti-aircraft (HAA) gun was developed in the late 1930s and was in production by 1938. The gun fired a 28 pound (12.7 kg) shell to an effective ceiling of 32,000 feet (9,750 metres). It was a very complex system which needed a considerable amount of supporting equipment, such as height-finders, predictors and fuse-setters. The gun was manufactured in Canada, starting in 1941, and production eventually reached 300 guns per month. In addition to the field forces, the gun was mounted in ports and other strategic locations in Canada to provide protection against air attack.

As of 1941 there was no complete Canadian heavy anti-aircraft regiment overseas. However, on the supposition that 3.7-inch AA equipment was likely to become available, it was decided to convert 2nd Canadian Medium Regiment to a heavy anti-aircraft regiment and despatch it overseas. Part of this unit arrived for training by Atlantic convoy in August 1941, with the rest to arrive shortly afterwards. However, 3.7-inch guns would not be available for the unit on the arrival of the main body and arrangements were made for this unit to proceed to British anti-aircraft training regiments.

In February 1942, the regiment was still the only Canadian heavy anti-aircraft unit overseas. It was employed by ADGB (Air Defence Great Britain) using guns supplied by the British anti-aircraft command. Of its total complement, only four guns 'belonged' to the unit, leaving a deficiency of 24 guns.

By November 1942, the situation with respect to 3.7-inch anti-aircraft guns had improved considerably and its establishment, now set at 24 guns, was complete. The regiment remained

A Wasp Mark II equipped Universal Carrier demonstrating its 'flaming' ability. This model had the flame tank internally mounted. Library & Archives Canada PA141867

A Canadian Ronson carrier. It's external tanks were preferred by CMHQ. Ultimately the Wasp 2c was chosen for operational use by the Canadian Army. Library & Archives Canada e002344198

the only Canadian HAA unit in the United Kingdom throughout the war, and proceeded to the continent with First Canadian Army. There, like other anti-aircraft weapons, the 3.7-inch gun was used mainly in a ground role, firing airburst with considerable effect.

Rocket Projectors

Towards the end of 1944, a new weapon was introduced into operations by First Canadian Army. This was the "Land Mattress", known officially as the Projector, Rocket 3-inch, No. 8, Mark 1. The rockets were a scavenger's delight. The 60-pound high-explosive warhead came from the Royal Navy, the Royal Air Force donated the rocket motor, and the army handed over 600,000 Type 721 fuses which had been scrapped for safety reasons,

Front and rear views of a Canadian Ronson carrier. The model on the right of both images sports a custom protective cover. This item was not put into standard production. The formation sign and Arm of Service markings on these two carriers identify them as belonging to the 1st Battalion, RCE, 1st Canadian Corps Troops. Photos courtesy Barry Beldam.

although First Canadian Army found them acceptable. The fuses then had to be sorted to find the correct version. Canadian Base Workshops manufactured adapters so that the parts could be made into a usable rocket.

Twelve prototype 32-tube projectors were manufactured, and the battery fired its first operational trial at German open-topped anti-aircraft gun positions in Flushing on 1 November 1944. The results were successful and the battery supported most major attacks for the rest of the war. The "operational trials" were deemed complete at the end of 1944, and the continued existence of the rocket battery was approved. With deliveries of the production projectors beginning in February 1945. It was proposed to establish a rocket projector regiment in First Canadian Army consisting of three batteries, each of twelve 30-barrel projectors, although the full regiment was never formed. The unit consisted of a small permanent establishment of instructors and employed personnel of light anti-aircraft regiments for operations.

The Canadian-developed system was sufficiently successful that, in January 1945, 21st Army Group asked for the equipment and ammunition for one regiment per army to be used in the same way as the Canadians.

Flame Weapons

Various types of flame-throwing weapons were employed by the Canadian Army during the war. Flame throwers had been used by the British on an experimental scale since 1940, but the development of these weapons as standard equipment depended largely on Canadian inter-

Ronson flame-thrower equipped universal carrier demonstrating its fearsome ability. Tactical doctrine required that these be heavily supported by infantry units to both protect the carrier and to take advantage of the effect it had on the enemy, most of whom surrendered immediately. Photo courtesy Department of National Defence

est. Various experiments had been carried out by the Canadian Army in the United Kingdom and the first production on a large scale was in Canada.

The first type developed for use by the Canadian Army was the Canadian-made "Ronson". This weapon was mounted on a Universal Carrier. Orders had been placed in Canada for 1,000 of these for the Canadian Army Overseas. First deliveries were ready by the end of 1942. A British proposal to use 50 of them in the Middle East was opposed by the

Churchill (Mark VII) Crocodile flame-thrower. After trials with flame fuel carried in trailers and tested on Valentines in 1942, the British General Staff decided to standardise on a flame-throwing system actuated by gas (nitrogen) pressure. Design was finalised in 1943 with an initial order for 250 units featuring armour protection for the fuel trailer, and the Churchill Mark IV was selected as the operating vehicle. In October 1943 the Churchill Mark VII was chosen instead. Fuel passed along the belly via a "link" with the trailer, and the projector replaced the hull machine gun. Late production Churchill Mark VIIs were all built for speedy adaptation to the Crocodile role as required. The vehicle's main armament could still be used. The range of the Crocodile was 80 to 120 yards in 80 one-second bursts from a full trailer. When empty or hit, the trailer could be jettisoned. These vehicles were used by especially trained regiments of the Royal Armoured Corps, all of whom were brigaded within the British 79th Armoured Division, and supported the Canadian Army at various times in Northwest Europe 1944-45. Library & Archives Canada e002505439

Canadian Army Commander on the grounds that surprise would be sacrificed by their use on a small scale. The proposal was abandoned. It is worth noting that none was yet available from British production. By the end of July 1943, the Canadian Army Overseas had received 818 Ronsons from Canada while the rest of the order had been lost at sea. Supplies were not adequate, but it remained to determine the method of their employment. Training was carried out at the Canadian Training School with personnel of reconnaissance regiments. The intention was to employ a maximum of 300 at one time using three reconnaissance regiments to operate 100 each. It was intended to hold these equipments in ordnance depots in the operational theatre for issue as required for specific operations.

The principal difficulty noticed in the Ronson was that of maintenance. There was some suggestion in March 1943 of adopting a new type, but the Army Commander expressed himself satisfied with the Ronson. The only alternative available at that time was the British Wasp Mark 1 which was not considered as good. In October 1943, however, the Canadian Army decided that the British Wasp Mark 2 Flame Thrower, just coming into production, was superior to the Ronson and placed an order for 500 of these with the War Office. The Wasp Mark 2 was a more advanced design using the principles established by the development of the

An Infantry Tank, Churchill, Mark I of the 11th Army Tank Battalion (The Ontario Regiment (Tank)), 1st Canadian Army Tank Brigade, being inspected by General McNaughton (hatless) and Brigadier Worthington (left), July, 1941. Library & Archives Canada e002344115

Infantry Tank, Matilda Mark II - this type was the first tank issued to the 12th Army Tank Battalion (The Three Rivers Regiment (Tank)), and the 14th Army Tank Battalion (The Calgary Regiment (Tank)), (1st Canadian Army Tank Brigade), upon their arrival in the United Kingdom, the 11th Army Tank Regiment (The Ontario Regiment (Tank)), being issued with the Infantry Tank, Churchill Mark I. Library & Archives Canada e002505510

Barracuda, a joint project of the Canadian Petroleum Experimental unit in conjunction with the British Petroleum Warfare Department. It was originally mounted, like the Ronson, on a Universal Carrier and had two fuel tanks mounted inside the carrier. The Canadian Army preferred to have one external tank in view of the difficulty of maintenance with an internal tank.

Infantry Tank, Churchill Mark IV, which saw service with the 1st Canadian Army Tank Brigade from mid-1942 ubtil replaced with Ram IIs in March 1943. This was the only Canadianb formation to be equipped with any Mark of the Churchill. Library & Archives Canada

Infantry Tank, Churchill Mark II of Headquarters Squadron, 14th Army Tank Regiment (The Calgary Regiment (Tank)), 1st Canadian Army Tank Brigade. Library & Archives Canada e002505518

The production of Wasps was proceeding rather slowly and the modified Canadian version, known as Wasp Mark 2c, was accordingly delayed. Meanwhile, Ronsons were used for training and it was intended to use them for operations if the Wasps were not available. They were not used in North-West Europe and the Canadian Army found itself with some 700 Ronsons available for disposal.

The policy in 21st Army Group was for the Wasps to be used by infantry carrier platoons as an occasional weapon, and stocks held in advanced ordnance depots were to be available on

Churchill Mark IIIs of "C" Squadron, 14th Army Tank Regiment (The Calgary Regiment (Tank)), pictured on parade, July 1942. Library & Archives Canada PA116274

Infantry Tank, Churchill Mark II of Regimental Headquarters, 11th Army Tank Regiment (The Ontario Regiment (Tank)), undergoing training with a Landing Craft, Tank (LCT). Library & Archives Canada e002505508

seven days' notice. Training was given to personnel of infantry carrier platoons at the Canadian Training School on the British Wasp 2. The first Wasp 2c was ready in June 1944. On the scale of eight flame throwers for each infantry divisional reconnaissance regiment, motor battalion and infantry battalion, First Canadian Army had a requirement of 192. Up to the latter part of November, First Canadian Army had received 134 Wasp 2 and 73 Wasp 2c. In keeping with

US-built Light Tank, M5A1 Stuart Mark VI tank of either the Reconnaissance Troop, Headquarters Squadron, 21st Canadian Armoured Regiment (The Governor General's Foot Guards), 4th Canadian Armoured Division or of the Reconnaissance Troop, Headquarters Squadron, 2nd Canadian Armoured Regiment (Lord Strathcona's Horse (Royal Canadian)), 5th Canadian Armoured Division. Library & Archives Canada PA113170

US-built M2A4 Light Tank. A limited number were issued to Canadian Armoured Brigades. Library & Archives Canada

Canadian policy, efforts were being made to have First Canadian Army equipped entirely with the Canadian model.

Early in 1944 another, and more powerful, flame-thrower was coming into production in

M3 Stuart Mark I light tank of "C" Squadron, 5th Armoured Regiment (8th Princess Louise's (New Brunswick) Hussars), 2nd Canadian Armoured Brigade, 5th Canadian Armoured Division, Aldershot, England, May 1942. Library & Archives Canada e002343947

A US-built M3A3 (Stuart) tank, nick-named a 'Honey'. This example has been de-turreted and is in use by 5th Armoured Regiment (8th Princess Louise's (New Brunswick) Hussars), 5th Armoured Brigade, 5th Canadian Armoured Division, Italy, 1944. Library & Archives Canada

US-built M3 medium tank (Lee). The exterior fuel tank was a common Canadian modification along with extra stowage boxes and sand chutes in the suspension. Library & Archives Canada

US-built light tank, M5A1 Stuart Mark VI tanks of the Reconnaissance Troop, Headquarters Squadron, 2nd Armoured Regiment (Lord Strathcona's Horse (Royal Canadians)), (5th Canadian Armoured Division), Eelde airfield, The Netherlands, 23 May 1945. Library & Archives Canada PA137911

M3 medium tank (Lee) on a tank firing range somewhere in England. The circular marking on the turret identifies this vehicle as belonging to "C" Squadron of a regiment. The colour of the circle would further identify the regiment's seniority within a brigade. Library & Archives Canada e002505520

An M3 Lee medium tank (Lee) in Canadian service somewhere in England. Library & Archives Canada e002343940

the United Kingdom. This was the Crocodile, a flame-throwing apparatus usable in all Churchill VII tanks. The Crocodile was a direct descendant of the Oke flame-throwing tank built on the Churchill Mark II. The Churchill Oke was developed by the Petroleum Warfare Department in 1942, in time for a flame-throwing vehicle to be tested under combat conditions

Due to shortages, the first 80 Ram II tanks were delivered to England without their guns or gun-mounts. One of these tanks is shown rolling past Lt-Gen Crerar, Senior Canadian Officer who takes the salute. Library & Archives Canada e002343930

on the Dieppe raid. The Oke combined a Ronson-style flame-thrower and the Churchill's jettisonable fuel container, fitted to most tanks at this time, instead of the trailer that was being used for this purpose on other tank mounted flame-thrower designs. The jettison fuel container was connected up to a projector and fitted to a Churchill tank where it was operated by a crew member from the hull machine-gunner's seat. Three of these vehicles took part on the Dieppe Raid, but all three were destroyed before they got into action. The effective range of the Oke flame-thrower was 40 to 50 yards. Other types of tanks and other marks of the Churchill could not be used for Crocodiles.

The fuel was carried in an armoured wheeled container pulled by the tank and the flame gun was mounted in place of the hull Besa 7.92-mm machine gun. These were issued on the scale of one per troop in each armoured regiment. First Canadian Army had a requirement for 100 Crocodiles, but they would have to be adapted to fit Sherman tanks. The American Army intended to use Crocodiles on the Sherman I which would also fit the Sherman III and V versions used by the Canadian Army. No production was started on the Canadian requirement owing to the prior claims of the Churchill model. At the end of August the requirement was cancelled. The American Army had abandoned the idea of using Crocodiles on Shermans, leaving the Canadians as the only potential users. The Canadian forces in North-West Europe frequently had the support of Crocodiles provided by the British 79th Armoured Division.

To meet the lack of Crocodiles, First Canadian Army in August requested to have a Wasp 2 mounted in a Ram Armoured Personnel Carrier. By December 1944 First Canadian Army had received 36 of these. These equipments, known as 'Cougars' and later 'Badgers', were issued for trial purposes on the scale of six to each armoured regiment of 2nd Canadian Armoured Brigade and six to the motor battalion of 4th Canadian Armoured Division.

Another flame-thrower employed during the war was the Lifebuoy, Mark 2. This was a portable weapon operated by one man. It was held in ordnance depots for issue as required for operations. It was intended for use particularly in house or wood clearing or in street fighting where the Wasp and Crocodile could not be manoeuvred. It was employed occasionally by the Canadian forces. In practice, it was found to be too heavy and to require considerable maintenance, and there was a lack of trained operators. A few Lifebuoys were held by Canadian

A Sherman Vc "Firefly" with 17-pounder of the 27th Armoured Regiment (The Sherbrooke Fusilier Regiment), 2nd Canadian Armoured Brigade, moving into action near Buron, 7-11 June 1944. Library & Archives Canada PA131391

forces in the United Kingdom for training at the Canadian Training School. There was a shortage of these equipments for the Canadian Forces in Italy.

Churchill Tank

Tanks became an equipment concern for the Canadian Army Overseas with the arrival of 1st Canadian Army Tank Brigade in June 1941. This formation began to receive tanks on a training scale immediately upon arrival. By July 1941 it had a total of eighteen "I" (Infantry) tanks Mark IIA (Matildas), four "I" tanks Mark IV (Churchills) with another temporarily immovable on a flat-car at Lavington station in Wiltshire, and two cruiser tanks. By September 1941 it had a total of 62 tanks of all types, against an establishment of 178.

The Brigade was to be equipped with two battalions of Matilda tanks and one battalion of the new Churchill tank - this latter was to be issued to the 11th Canadian Army Tank Regiment (Ontario Regiment). The Churchill tank apparently did not impress British officers and many difficulties were encountered in its use. Brigadier (later Major-General) Worthington, who then commanded 1st Canadian Army Tank Brigade, nevertheless had faith in this heavy and powerful weapon and asked that his whole brigade be equipped with it. The War Office readily agreed to this and the Matildas were replaced by Churchills shortly afterwards. By February 1942, the 1st Canadian Army Tank Brigade was almost completely equipped. It held four Cruiser tanks (Covenanter type of an early Mark) for its Headquarters Squadron, and 160 Churchills of an establishment of 174. This was in addition to various scout cars and carriers.

In May 1942, the 1st Canadian Army Tank Brigade's Heavy Support Company received nine "Gun Carrier, Churchill, Mark I", These were Churchills with 3-inch, 20-cwt anti-aircraft guns and were used in an anti-tank role until such time as the 17-pounder anti-tank gun was ready for issue. The unit was broken up in February 1943.

Sherman flail tank. This 'funny' was developed to clear a path through minefields. The drum rotated, causing the chains to churn up the earth and detonate buried mines. The whole mechanism was spring-mounted to help absorb detonations. Library & Archives Canada PA131366

Throughout the spring and summer of 1942 the Churchill was used by 1st Canadian Army Tank Brigade in a series of exercises, the most important being "BEAVER III", on 23 April 1943. The difficulties experienced with this vehicle in previous training became most apparent in this exercise, and "A Brief History of 1st Canadian Army Tank Brigade", prepared by the headquarters of that formation, contained the following comments on the Churchill's performance. *"The Brigade commenced the exercise with 139 Churchills of varying quality. During the five days that the brigade was on the move, an average of 135 miles was put on each tank: 119 of these tanks were reported as ... casualties, that is at one time or another during the five days almost 90 percent of the tanks were reported "off the road" with either a major or a minor breakdown. The terms "breakdown", "stragglers", "limpers", and several others describing these tank casualties became household words, and "oil seals", "gear boxes", "clutch", "amal drives", and "stuck starter" were added to the vocabulary of a great many, these being the major faults encountered."*

In an attempt to overcome the apparent lack of mechanical reliability encountered by the users of the Churchill, the War Office detailed a system of re-working this vehicle, and many of these modified tanks were issued to 1st Canadian Army Tank Brigade following Exercise "BEAVER III". Some of these re-worked models participated in the Dieppe operation. The evidence of Major C.E. Page, after his repatriation from Germany, substantiated the statement earlier made in The (London) Times of 21 July 1941 *They were heavily armoured giving the maximum of protection to their crews ..."* Major Page commanded 'B' Squadron, 14th Canadian Army Tank Regiment, during the Dieppe operation. He stated that no Churchill tank was pierced by enemy fire, nor was any man wounded inside a tank, despite the fact that many were immobilized shortly after landing and consequently subjected to heavy enemy fire.

The Churchill, despite its effective protection, was not considered a success. Shortly before Dieppe, at a time when Canadian tank production plans were much under discussion, Lieutenant-General A.G.L. McNaughton despatched to the Vice Chief of the General Staff, a personal telegram containing an appreciation of the tank situation in 1st Canadian Army Tank Brigade. His proposals for future policy concerning the equipment of this Brigade were; *"...1 Canadian Army Tank Brigade [is] now equipped with Churchills on loan from War Office as*

Sherman Crab II (Flail) tank of the 1st Lothian and Border Horse, 30th Armoured Brigade, 79th (British) Armoured Division, in support of the 5th Canadian Armoured Division, passing through Putten, The Netherlands, 18 April, 1945. Library & Archives Canada

an interim arrangement pending supply of Ram tanks from Canada. It has been intended to retain Churchills until after 4th and 5th Canadian Armoured Divisions were equipped with Rams. 1st Canadian Army Tank Brigade has been in the UK for 13 months and is fully trained to very high efficiency. Against establishment of 178, tank state today is as follows: original model 121, re-worked tanks 37, new model 24, total 182. Of these 63 are non-runners including 24 for which spares are not available. Despite every effort it has proved impossible to keep the Churchills in running order, first due to certain vital defects in original design and second due to impossibility of obtaining spare parts. I am informed by HQ Royal Armoured Corps that situation will continue to deteriorate for some months before we can expect increased availability of spares and re-worked or new model tanks. Another serious factor is that personnel have lost confidence in mechanical reliability of Churchill and on this account I am most anxious to replace them with re-worked or new model and then and as soon as possible, with Ram II or preferably M4 (Sherman)."

Throughout the remainder of 1942 and the first three months of 1943, the newer model Churchill, Mark IV, was issued to the Brigade and the older tanks, which *"had covered many miles and were generally in very poor condition"* were withdrawn. Although the Brigade was not committed to action until the last day in Exercise "SPARTAN", it did make several long marches and *"the improved Churchill with a better organized Ordnance Company made all the difference"*.

In March 1943, the Churchills were withdrawn from 1st Canadian Army Tank Brigade, the only Canadian formation which had possessed them, and replaced with Ram IIs.

Sherman V, ARV Mark I - an armoured recovery vehicle based on the M4A4 chassis used in armoured and armoured reconnaissance regiments, as well has by RCOC/RCEME units for the recovery of armoured fighting vehicles. Library & Archives Canada PA116535

Another Canadian-manned Sherman V, ARV Mark I. This one is kitted-out slightly differently from the one above. Library & Archives Canada e002505437

A Sherman 'Command' tank of Headquarters, 2nd Canadian Armoured Brigade, during the offensive south of Caen, Normandy, 1944. Note the added welded tank tracks and sandbags on the front of the hull for added protection from enemy anti-tank fire. Library & Archives Canada PA114062

Light Tanks

Pending the availability of Ram Cruisers, American tanks had been provided by the War Office. By February 1942, 5th Canadian Armoured Division held 40 "General Lee" Cruiser tanks and five "General Stuart" Light tanks. As of that date no other Canadian formation used these tanks.

In October 1943, the decision was taken to use light tanks in place of carriers in armoured regiments following the British practice. At about the same time the War Office adopted a new establishment for armoured reconnaissance regiments which called for 30 light tanks per regiment. This was adopted by the Canadian Army early in November. The War Office undertook to supply Stuart V light tanks (Light M3A3) to cover Canadian requirements. Unfortunately, the supply situation was poor and some of the Stuarts received were in bad condition. This situation showed little improvement during the next five months. At the end of May 1944, the Canadian Army held only 36 Stuart tanks against an establishment of 77 in six armoured regiments and one armoured reconnaissance regiment (later re-organized as an armoured regiment), and the unsatisfactory condition of many of those delivered still prevailed. The 2nd Canadian Armoured Brigade was, however, almost completely equipped. By the end of June, before they had seen action, 4th Canadian Armoured Division had also received all but two of their Stuart tanks. Early in August it was possible to release nine Stuarts surplus to field requirements to Canadian Reinforcement Units.

A proposal was made to replace some or all of the light tanks in the Canadian Army with scout cars. Although this was not carried out, in October 1942 steps were taken to replace the Stuart V with some other light tank. A number of American M24 light tanks had been allotted to the United Kingdom on the understanding that they would not be issued until the American troops had them. They would be issued to Canadian formations through 21st Army Group. It mounted a 75-mm gun as compared with the 37-mm carried by the Stuart V. Meanwhile, a pol-

A damaged Sherman DD tank of the 10th Canadian Armoured Regiment (The Fort Garry Horse), 2nd Canadian Armoured Brigade, mired on the beach near Bernieres-sur-Mer, 6 June, 1944. Library & Archives Canada PA132897

icy was adopted of replacing all Stuart V Tanks with Stuart VI's (M5 and M5A1). Early in November First Canadian Army reported that this replacement was nearing completion. At the end of December the Canadian Army held 76 Stuart VI tanks against an establishment of 77. The deficiency was covered by three Stuart V tanks still held, but no M24 tanks had yet been issued.

Although the Stuart V was not particularly satisfactory in North-West Europe. In that theatre the Canadians employed them principally with the turrets removed and a .50 Browning machine gun mounted on top. Known as "Honey" tanks the Stuarts were frequently mentioned as giving good service.

In addition to the Lees used by the 5th Canadian Armoured Division, a certain number of Grants were subsequently issued. The Equipment State for 31 August 1942 shows 23 of these tanks on issue. These tanks were also used by the 4th Canadian Armoured Division after its arrival in the United Kingdom. By November 1943, only one Grant remained on charge. There were also seven obsolete M2A4 tanks in stock. Stuarts were, at the time, once more required as light tanks for certain units.

Ram

1st Canadian Army Tank Brigade was not the first formation to receive the Ram cruiser tank. 5th Canadian Armoured Division had received some Rams in March 1942. The decision to equip 5th Canadian Armoured Division with Rams had been made as early as 1941, and the production target of the tank industry in Canada for 1942 was 1,155 Ram tanks. It would appear therefore that hopes had been entertained of completing the war establishment of the division before December of that year. However, the numbers of Ram tanks arriving in the United Kingdom during 1942 were not sufficient. The establishment of an armoured division was 340 Cruiser tanks. Information had been received by CMHQ that 40 Ram I tanks (2-pounder gun) and 300 Ram II tanks (6-pounder gun) would be shipped from first production to equip the 5th Canadian Armoured Division. Of these, the first 80 Ram II tanks were shipped without guns or gun mounts and these were shipped later for installation in the United Kingdom.

A Sherman blowing the waterproof sealant. All tanks used in the initial asault were waterproofed with a fabric tape and rubber compound. This was explosively removed by a length of pre-positioned detonating cord which was wired into the tank's headlights. Library & Archives Canada

The failure of Canadian tank production to meet the demands of the army in 1942 can be attributed to two factors. The first was directly associated with the actual manufacture of tanks. Canadian tank industry leaned very heavily upon the United States for the supply of essential components, the Ram being based on the American M3 chassis. When it was learned, in October 1942, that the United States had reduced the allocation of transmissions by 40 percent, it became apparent that delivery of the first 1,000 Rams would not be complete until mid-December. In the same period, Montreal Locomotive Works, the Canadian manufacturer of the Ram, was called upon to produce 25-pounder self-propelled guns on the Ram chassis - the Sexton. In effect, this meant that for each self-propelled (SP) gun produced, one Ram tank fewer would leave the assembly line. The second cause was shortage of shipping. Even though the total number of tanks produced was not as great as originally planned, many of those earmarked for export to the United Kingdom were left at the seaboard owing to the non-availability of cargo space. Thanks to the success of anti-submarine operations and other causes, the shipping situation improved early in 1943, and in consequence a greater proportion of Canadian output was shipped overseas from January onwards. The result was that by 1 June 1943, 1,147 Rams were in the hands of the Canadian Army in the United Kingdom. These were sufficient to complete the establishment of Canadian armoured formations overseas.

It was not enough that the Canadian Army should be provided with a complete establishment of tanks. If the Canadian armoured formations were to acquit themselves successfully in operations, it was essential that their vehicles be battle worthy. The Ram II did not fulfil this requirement. In particular, its armament was, by 1943, considered inadequate. Although the 6-pounder was a much heavier weapon than that carried by British tanks only a few months before, military opinion now tended to favour heavier guns such as the 75-mm gun mounted by the US Sherman tank, which had been so successful in North Africa. In addition, a great many mechanical and other modifications to the Ram had been found necessary. An undated memorandum by Colonel H.A. Guy, CMHQ, evidently written in March 1943, listed a total of 113 major and minor modifications which had been suggested, and called for the General Staff to establish a priority list to deal with these. The list of modifications included a recommendation that increased protection at several points be provided.

The Grizzly cruiser tank was a Canadian-built version of the M4 pattern, manufactured by Montreal Locomotive Works. No Grizzly ever made it into combat. Most were kept in storage in Canada and used postwar for training and, finally, as range targets. Army Technical Development Board Report.

In order to equip the Canadian armoured formations with battle-worthy tanks, it was necessary either to produce a new tank (a very long process) or to undertake radical modifications of the Ram, including arming it with a 75-mm gun. On 3 June 1943, a meeting at HQ First Canadian Army considered the situation. General McNaughton said that nothing in any report which had been brought to his attention indicated that the Ram II was not capable of being modified so as to become a completely battle-worthy tank. He said that of the important modifications, the provision of a 75-mm gun was essential, and was of high priority. The meeting unanimously agreed. National Defence Headquarters was notified, after this meeting, that 600 of the available Rams would be reserved for operational use by an armoured division or tank brigade if required before the main target date tentatively set for First Canadian Army's probable employment. These tanks *"would have all necessary modifications including 75-mm guns completed earliest possible date and after sufficient running in would be held in stock on immediate availability basis"*. The balance, approximately 1,000 Rams, would be used for training requirements of Canadian Armoured Corps units and reinforcements. They *"would have only minimum modifications"*, but for gunnery training purposes 100 of them would be armed with 75-mm guns and allotted to units and firing ranges. NDHQ was further told: "...*for full scale equipping of First Canadian Army against main target date, McNaughton desires M4 type tank armed with 75-mm gun and powered with either Wright radial or Chrysler engine* ..."

Subsequently, however, it was decided to equip the Canadian Army Overseas completely with M4 (Sherman) tanks. In consequence it became unnecessary to make the alterations to Rams detailed above.

It should be noted that an important element in the Ram situation in 1943 was the difficulty of implementing the modifications. General McNaughton, in a telegram to the Chief of the General Staff (CGS) dated 29 May 1943, observed that *"extensive modification requirements ... have swamped our BOW"* (Base Ordnance Workshop) and referred to *"inability to secure contractors to help out."* In view of this and of the many advantages of the M4 tank,

Variant of the Loyd light anti-aircraft tank, Mark I, mounting twin 7.92mm Besa machine guns. Library & Archives Canada e002055366

General McNaughton wrote " ... *It seems likely that we will be forced to conclusion that M4 should be provided for any formations required for operations, and that Ram should now be reserved for training only ...*"

A detailed comparison of the Ram and the Sherman was made by a committee assembled by Brigadier R.A. Wyman, commanding 1st Canadian Army Tank Brigade, a formation that used both vehicles. Its report made the point that *"The general suitability of the Ram II does NOT meet the requirements of a first line operational tank. Its disadvantages are far more numerous than its advantages in comparison with the Sherman ..."* Among the points made were the superiority of the 75-mm gun to the 6-pounder, and the facts that *"The Sherman ... is superior on rough going or hill climbing and better for general cross country work"* and that *"the Sherman is an easier and much less fatiguing tank to handle than the Ram"*. The Sherman's traversing mechanism and firing controls were both described as *"more satisfactory"*, while *"From experience, it is estimated that the vision from a Sherman is infinitely better than from a Ram"*. Both protection and escape arrangements in the Sherman were considered to be superior. The point was also made that whereas in the Sherman *"the controls are very convenient ... in the Ram the controls are so situated that the driver becomes very fatigued after driving for a short distance"*.

Although the Ram was not to be employed in operations as a cruiser tank, it had played an important role in training the armoured formations of the Canadian Army Overseas. Moreover, nearly 500 Rams were made available to the War Office for use as Assault Vehicles, Royal Engineers (AVRE). Others were converted to various roles, including recovery, ammunition carrier and personnel carrier.

The total production of Rams was 1,949, of which 277 were retained in Canada for training and other purposes. The last of the overseas assignment of 1,671 had been shipped by December 1945. Of these 104 were lost at sea and only 1,567 were actually received.

Crusader III, Mark II anti-aircraft tank - armed with twin 20mm Oerlikon anti-aircraft cannon - use was limited to the anti-aircraft troop of the headquarters squadron of an armoured regiment and of an armoured reconnaissance regiment, there were also two on the strength of an armoured brigade head-quarters. Library & Archives Canada PA193591

Sherman

The American M4 Medium Tank (Sherman) made its appearance in the British Eighth Army in the Middle East in the late summer of 1942. Mr. Churchill had described the manner in which *"the admirable Shermans"* were acquired. He was in Washington with President Roosevelt in June of that year, when news came of the British reverse in Libya and the loss of Tobruk: *"... Nothing could have exceeded the delicacy and kindness of our American friends and Allies ... Their very best tanks - the Shermans - were just coming out of the factories. The first batch had been newly placed in the hands of the divisions who had been waiting for them and looking forward to receiving them. The President took a large number of these tanks back from the troops to whom they had just been given. They were placed on board ship in the early days of July, and they sailed direct to Suez, under American escort for a considerable part of the voyage."*

The 'General Sherman', as it was originally termed by the British, first appeared in the Canadian Army in the spring of 1943, when this tank was issued to the 1st Canadian Army Tank Brigade (later reorganised as 1st Canadian Armoured Brigade). This formation had had Rams for only a few weeks when it was decided to despatch it to the Mediterranean, along with the 1st Canadian Infantry Division, for Operation HUSKY (the invasion of Sicily). With this in view, the brigade was moved to Scotland for training late in April, where it was equipped with Shermans. It used these tanks throughout the campaign in Sicily.

The Sherman was a 30-ton tank of United States design and manufacture, a direct development of the M3 types known by the British Army as the 'General Lee' and 'General Grant'. It represented a fusion of the best features of US and British tank practices, American mechanical excellence and heavy armament, but with the latter mounted as in British armoured fighting vehicles (AFV). The Grant and Lee were armed with a 75-mm gun, mounted on the starboard side of the hull, where it had extremely limited traverse and could not be used when the tank was in hull-down position (positioned in such a way that only the turret is exposed) . The Sherman mounted the 75-mm gun in the turret after the British manner. The chassis was an improved version of the M3, while the hull was much more streamlined and the profile more

Only three Skink anti-aircraft prototypes, and an additional eight turret kits, were built. One example deployed to Northwest Europe for operational trials. A special unit, 1st Canadian Tank Demonstration Unit, was created for these trials. The Skinks were built on the Grizzly chassis. MilArt photo archives

compact. The War Office later proposed to install a 17-pounder gun in a proportion of Sherman tanks, i.e., 15 per armoured regiment.

These tanks played a material part in the Eighth Army's great victory at El Alamein in October 1942, as well as in the subsequent Allied successes in the Mediterranean theatre generally. At the time it appeared to be a matter of common consent that the Sherman was the most generally satisfactory tank produced to date by the Allies. Reports received of its employment by 12th Canadian Army Tank Regiment in the Sicilian campaign indicated that the unit was well satisfied with its performance, and particularly with its 75-mm gun. The 5th Canadian Armoured Division took no heavy equipment to the Mediterranean theatre and were issued Sherman tanks in Italy.

As the war progressed, the policy of using Sherman tanks in the Canadian Army remained in effect. The majority of those required for operations had been received from the British by the end of 1943, but apart from a few for training purposes they were not issued to units until April and May 1944. The policy was to use Ram tanks for training both in field units and reinforcement units. Besides saving the Shermans for operations, this permitted a considerable amount of necessary work in modifying the Shermans to be accomplished before they were issued. This work was not completed until April 1944. Canadian units were fully equipped with Shermans before the start of operations in North-West Europe. Except for a slight deficiency in November the Canadian Army was kept fully supplied.

The model chosen for use in the Canadian Army was the Sherman V (M4A4), but due to the supply situation it was found necessary to equip the assault brigades with the Sherman III (M4A2) throughout 21st Army Group. The Sherman III had a Twin Diesel engine (GM6-71) not used in any of the other marks of this tank but which was used in the 3-inch M10 self-propelled anti-tank gun.

The policy of mounting 17-pounder guns in 15 tanks per armoured regiment proved dif-

A Ram Badger flame-thrower of the Badger Troop, Headquarters, 5th Canadian Armoured Brigade, Putten, The Netherlands, 18 April 1945. Library & Archives Canada PA131031

M3A3 Stuart V de-turreted light tanks of, on the left "B" Troop Commander, and on the right "C" Troop Commander, 16th Anti-Tank Battery, 4th Anti-Tank Regiment, 5th Canadian Armoured Division, Eelde, The Netherlands, 23rd May 1945. Library & Archives Canada PA138031

ficult to attain. The gun used was the Mark 4 and later the Mark 7, which differed considerably from the Mark 1 used in the anti-tank regiments. Most were mounted in Sherman V tanks and when these were no longer available it was decided to use the Sherman I as well. These tanks, known as Sherman Vc and Ic once upgunned, continued to remain in short supply until near the end of 1944, and it was necessary to use tanks mounting the normal 75-mm gun to cover the deficiencies.

Ram Kangaroo Armoured Personnel Carrier. These were based on the Canadian-built Ram II Crusier Tank. These vehicles were also used by Armoured Regiments as armoured ammunition carriers to carry ammunition to tanks in action. Library & Archives Canada PA137744

A Priest Kangaroo of the 4th Canadian Armoured Division in Delden, 4 April 1945. This configuartion still retains the additonal armour added to the sides when it was a self-propelled gun. It appears to have been commandeered as an armoured command post. Library & Archives Canada PA131627

For the actual assault landing in Normandy in June, some special tanks were issued to 6th and 10th Canadian Armoured Regiments of the 2nd Canadian Armoured Brigade. These were Sherman Duplex Drive (DD) or "swimming tanks", fitted with two propellers. Two types of tanks were adapted to the DD configuration, the Sherman V and Valentine. However, the Canadian army used only the Sherman and each regiment received 38 of these, to be replaced

Cruiser Tank, Crusader Mark III - the final production version with 6-pounder replacing the 2-pounder gun, and increased armour on the hull and turret. In production from May to July 1942, only 144 were produced and were limited to training purposes. Library & Archives Canada e002505353

as soon as possible after the landing with ordinary Sherman III tanks. Due to the weather, less than half of the DD tanks swam to shore, the others being landed in the same way as other tanks. Of those that did swim, a large number were drowned or otherwise lost. But the rest reached shore and remained as planned - hull down in the water and providing the only close support for the infantry in the early stages of the invasion.

The Canadian forces in Italy were also equipped with Sherman tanks. Reports both from there and from North-West Europe spoke favourably of the performance of the Sherman, though desires were expressed for more of those mounting 17-pounder guns. Both the 75-mm and the 17-pounder guns were reported as satisfactory, but the shortage during most of the year of the 17-pounders resulted in a lack of training of reinforcements in that weapon. Towards the end of 1944 Shermans mounting 105-mm (howitzer) guns were issued to the Canadian Forces in Italy on a scale of six per regiment.

Grizzly

During 1942, plans were made for the production in Canada of a tank which would be an improvement on the Ram and would form, it was hoped, part of a co-ordinated North American production programme. The vehicle finally decided on, and christened the "Grizzly", was an M4-type tank using a 75-mm gun or the higher velocity T13 (76-mm). This was in accordance with the views of General McNaughton, expressed on 2 August 1943. The Grizzly was built by Montreal Locomotive Works after completion of the Ram II and Ram OP/Command contracts. With a view to producing this tank, the Canadian Government, by 20 January 1943, placed orders for 80 percent of the materials required for the production of 1,200 Grizzlies by February 1944.

Following considerable discussion between the British and Canadian authorities during the summer of 1943, the War Office agreed to equip all four existing Canadian armoured brigades and the two armoured reconnaissance regiments with Shermans from US production. It was also agreed that Canadian tank production was to be divided between the Grizzly, which was to be produced at the rate of about 50 per month to a total of 250, and the self-propelled 25-pounder (Sexton, see below), production of which would be increased to 150 per month.

A Churchill ARK Mark II (Italian Pattern). This was an "Armoured Ramp Carrier" comprising an unturreted tank with a timbered trackway on the hull top. Library & Archives Canada e002505170.

A 'scissor' bridge laid by a Valentine bridge-laying tank being traversed by a Humber Mark III, Light Reconnaissance Car. Library & Archives Canada PA167223.

Before the Grizzlies were completed, however, the need for anti-aircraft tanks led to the decision to convert these vehicles to an anti-aircraft role by substituting the "Skink" (*see below*) anti-aircraft turret. The 250 Grizzly chassis originally called for were in production, but as original requirements changed to a need for 360 Skinks, an additional 110 chassis had to be produced and contracts were issued for these. However, changing fortunes reduced the need for the Grizzly and only 188 were manufactured.

Universal Carriers of the Le Regiment de Maisonneuve, the Royal Regiment of Canada and the Essex Scottish Regiment, England, May 1941. Library & Archives Canada e002344114

Specialty Tanks

Armoured Vehicle Royal Engineer (AVRE) were tanks converted for various special uses by the Engineers such as carrying 'fascines' to fill short gaps. They were also operated in Northwest Europe by 79th (British) Armoured Division. The tanks used were Churchills. AVREs were used in support of Canadian troops both in Italy and Northwest Europe.

Although Ram tanks were not used in action as cruiser tanks, a variety of other uses was found for them. As already mentioned they played an important part in training and continued to be used in Canadian Reinforcement Units for training. At the end of 1943, 446 Rams were transferred to the British for conversion into Armoured Vehicles Royal Engineer (AVRE). These were intended to replace the Churchill in this role. About half the tanks turned over to the British were converted into armoured recovery vehicles (ARV). Another third were turned into Kangaroos (*see below*), an armoured personnel carrier, a variant developed by the Canadians. The remainder were used for various miscellaneous jobs. Of those still in Canadian hands, about 300 were converted into armoured personnel and armoured ammunition carriers for First Canadian Army. The War Office accepted a further 330 of these for use in Italy. Thirty-six of these Ram carriers were equipped with Wasp flame throwers, while another 59 Rams were converted into towers for 17-pounder anti-tank guns. They were used by 6th Canadian Anti-tank Regiment of 2nd Canadian Corps. Still others were used as observation posts (OP) for artillery units. At the end of November, First Canadian Army held a total of 128 Rams of various types, not counting Sextons. The British were expected to take over all but 250 of the remaining Rams; these 250 would be used for training in the United Kingdom.

Armoured recovery vehicles used by the Canadians were intended to be Rams. About 50 of these tanks were converted for this purpose. But, early in 1944, it was decided to use Shermans in this role. By the end of May, Canadian units in First Canadian Army were almost completely equipped with Sherman ARVs. Shermans were also used as recovery tanks by the Canadians in Italy, where they were reported as working well.

At the end of 1943, the stated policy of the Canadian Army was to use British Crusader or Centaur tanks in the anti-aircraft role. The plan was to later replace these with Canadian

The ubiquitous Universal Carrier in its plainest appearance. Many variations of this vehicle were adopted throughout the war. Library & Archives Canada

A Universal Carrier (MMG) of a Medium Machine Gun platoon of the Saskatoon Light Infantry (MG), 1st Canadian Infantry Division. Library & Archives Canada PA189892

Loyd carrier of the Overseas Canadian Training School shown with Vickers Medium Machine Gun.
Library & Archives Canada e002344120

"Skink" anti-aircraft turrets mounted on Sherman tanks when they became available. Later it was planned to use complete tanks from Canada, i.e., Skink turrets mounted on the Grizzly chassis.

Canadians used two types of Crusader anti-aircraft tank. These mounted either a 40-mm Bofors gun (Mark I) or twin 20-mm Oerlikon guns (Mark II). The Canadian Army was slow in receiving the Crusader anti-aircraft tanks. A deficiency of 13 still existed in June, but the army was eventually completely equipped with these specialist vehicles. By June 1944, thanks to Allied air superiority, there was little need for anti-aircraft tanks. In August, the Crusaders were withdrawn from units of the First Canadian Army, though towards the end of the year some were being used in a ground role.

Bridge-laying tanks became a renewed requirement for the Canadian Army early in 1944. The total requirement was for six of these specialty vehicles, all of which were supplied from British sources. The tanks used were Valentines. In this connection a few Ram tanks were turned over to the British for special use with Bailey Beach Bridges. These were for "ship to shore" use - the Ram tank both supported the bridge and manoeuvred it into position from the shore end. The Canadians, on occasion, had the use of ARKs (Churchill tanks with bridging on top) supplied by the 79th (British) Armoured Division. In the Italian theatre, Shermans were modified to carry and launch the Bailey bridge. In North-West Europe, the idea was introduced of using tanks without modification to manoeuvre the Bailey mobile bridge to a gap.

In addition to these specialist tanks used by Canadian troops, various types of tanks, manned by British troops, were used from time to time in support of the Canadian forces. Flail tanks, sometimes known as "Crabs", were tanks fitted with whirling chains on the front and were used to breech minefields. They were operated by a special Flail Regiment of 79th

A T-16 carrier towing a 6-pounder anti-tank gun. The detachment had two carriers, one of which towed the gun and carried four members of the detachment (plus the driver). The second carried stores, ammunition, and the fifth crewman and another driver. In this case, the fifth member of the detachment is sitting on the rear deck. The sergeant standing beside the driver is the detachment commander. No. 4 gun, "N" Troop, 108th Anti-tank Battery, 2nd Anti-tank Regiment, (2nd Canadian Infantry Division). Library & Archives Canada PA191132

The turret of a Light Anti-aircraft tank Mark II - these turrets contained four 7.92mm Besa machine guns mounted abreast, This tank was issued on a scale of four per headquarters squadron of an armoured regiment for anti-aircraft protection, throughout 1941. Library & Archives Canada e00205522

A line-up of US-built T-16 Carriers, in Canadian service. The crews have augmented their firepower with the addition of pedestal-mounted US-made Browning .50 calibre heavy machine guns. Library & Archives Canada.

Loyd Carrier, Mark I. This was used as gun tractors for 6-pounder anti-tank guns, as a battery charger carrier and as a personnel carrier. Used extensively in England they were replaced by Canadian-made Universal Carriers before the invasion of Northwest Europe. Milart photo archives

US-made T-16 Carrier of The Lake Superior Regiment (Motor) (the Motor Battalion of the 4th Canadian Armoured Brigade, 4th Canadian Armoured Division) crossing the Afwalnings Canal, by raft, 7 April 1945. Library & Archives Canada PA167198

Truck, M14, 15-cwt, Half-track, which in Canadian service was used in a number of roles, including armoured personnel carrier in forward areas. With modifications it was used as a command vehicle, ambulance, ammunition carrier, gun tractor or as a GS load carrier. Library & Archives Canada

Truck, 15-cwt, Armoured (C15TA), whose primary role was as an armoured personnel carrier in forward areas. It could also be converted into an ammunition carrier, GS load carrier or into a two-berth ambulance. Library & Archives Canada PA115394

(British) Armoured Division using Sherman tanks. They were made available on several occasions to assist Canadian operations.

Skink

In an effort to provide adequate protection from air attack, and to offer greater firepower then that offered by the twin-barrelled Centaur, Canada developed a four-gun anti-aircraft turret. Termed the Skink, this weapon provided a quadruple 20-mm mount for Polstens. It was designed in Canada and built by the Montreal Locomotive Works. The total order was for 265 turret kits, of which 130 were for the British, while the remainder were to be mounted on Grizzly tanks for the Canadian Army. In February 1944 there was some discussion about reducing this order, but Canada was advised to continue the production. In April, Allied vulnerability to enemy air attack was considered to be substantially reduced and General Crerar advised 21st Army Group that anti-aircraft tanks would not be required. In view of the possible ill effect on the morale of those engaged in their production the contract was not cancelled until this was confirmed. Finally, when advice from Canada indicated sufficient work in producing the Sexton would make up for the cancelation of the Skink programme, it was agreed to cancel the whole order. The cost of this cancellation was estimated at $7,600,000 whereas to complete the order would have cost $14,025,000.

Three Skinks were actually completed - as well as eight turret kits - one of which was shipped to the United Kingdom for trials. In November, this one was called forward to Northwest Europe for operational trial. Ironically, as the Allied Air Forces had achieved almost total superiority of the skies in Europe, there was only one occasion on which an enemy aircraft came within range of the Skink - which was, at the time, unmanned. The Skink did perform admirably in a ground role where its four 20-mm guns were used with great success against enemy strong points.

US-built Staghound, T17E1, Armoured Car. These vehicles were used by armoured car regiments attached as Corps Troops. In 1st Canadian Corps this was the 1st Armoured Car Regiment (Royal Canadian Dragoons) and in the 2nd Canadian Corps it was the 18th Armoured Car Regiment (12th Manitoba Dragoons). Library & Archives Canada PA160728

This Staghound has been converted into an mobile armoured signals vehicle. The 37-mm gun has been removed to provide space for two No. 19 radio sets and other, additional signals equipment. The vehicle shown here belongs to "A" Squadron, 18th Armoured Car Regiment (12th Manitoba Dragoons). Library & Archives Canada PA130920

Staghound armoured car of 1st Troop, "A" Squadron, 18th Armoured Car Regiment (12th Manitoba Dragoons), (2nd Canadian Corps Troops), the Hochwald, Germany, 2 March, 1945. Library & Archives Canada PA144144

Carriers

During the summer of 1941 deliveries of Universal Carriers, in quantity, from Canadian sources greatly eased the shortage. As of the end of 31 August 1941, however, there was still a considerable deficiency. Units of the Canadian Corps had 519 carriers on hand against an establishment of 730. The situation had significantly improved when, by February 1942, the position of the Canadian formations was considerably superior to that of the British formations which held only 80 percent of their allotment. Part of the reason was that Canada used the Universal Carrier in lieu of the Universal Mortar Carrier and Observation Post (OP) Carrier, of which divisional establishments were 63 and 27 respectively. At the time, 1st Canadian Infantry Division held only six Universal Mortar Carriers and six OP Carriers while the other two infantry divisions had even fewer. On the other hand, British divisions were in a similar position but were without the Canadian surplus of Universal carriers.

1st Canadian Tank Brigade was complete in carriers with 33 Loyd Tracked Personnel Carriers and 13 Slave Battery Carriers (SBC). The latter was to carry large batteries which served to start tank-engines when cold. 5th Canadian Armoured Division had few carriers. While their divisional establishment was 90 Universal Scout, 14 Universal Bren and 15 Universal Mortar Carriers, in addition to 27 Slave Battery Carriers and 9 Armoured OP, as of 31 January 1942 the Division's holdings were limited to 15 SBCs.

By 31 August 1943, units of the Canadian Army Overseas were in possession of 2,359 Ford-produced Canadian Universal Carriers, compared to only 143 carriers of British manufacture.

By 1943 carriers shipped from Canada were equipped with the 95-horsepower motor, rather than the older 85-horsepower. But those of the latter type in the United Kingdom were not to be modified. Carriers modified to carry the 3-inch mortar were in good supply, and there was an overall surplus in the Canadian Army Overseas, although the majority of the vehicles, by 1 December 1943, had not been issued to the field units.

By the beginning of 1944 the number of British-made vehicles within Canadian units had

One of two personal 'chargers' used by Lt-General Guy Simonds, Corps Commander, 2 Canadian Corps. This Staghound has been modified by the removal of its turret and the addition of extra radio equipment. Library & Archives Canada PA116585

further decreased, with only 72 being held as compared with 3,888 Ford carriers at the end of February. The 3-inch mortar carriers, numbering 942, were all Canadian-made, as were 149 carriers fitted for Ronson flame throwers. Although in September, the Ford factory began the model change to the new Windsor carrier, by the end of the year a total of 8,901 Universal Carriers had been produced against Canadian orders of which 6,985 had been shipped overseas. In addition, 18,650 were produced and distributed to Allies through the British Ministry of Supply and the Canadian Mutual Aid Programme. Many of the Canadian Universal Carriers were the Mark I and the establishment called for use of the Mark II (Welsh Guards Stowage). To remedy this, a programme was undertaken in which 1,891 carriers were converted to the Mark II pattern in the United Kingdom before issue to units.

Supplies of the Universal Carrier and 3-inch Mortar Carriers were adequate throughout 1944, and expected deliveries covered estimated wastage up to the end of 1945. Special uses of the Universal Carrier were to mount the Wasp flame thrower and as a lightly armoured observation post. The need for carriers adapted for the Vickers Medium Machine Gun was realised in 1944 and these were obtained from British sources. Late in the war a number of units mounted US .50 calibre heavy machine guns to provide additional firepower.

Besides the standard Universal Carrier the Canadian Army used the American T16 carrier, also a Ford vehicle. This was a later type of the Universal Carrier with increased stowage space and load carrying capacity. It was adopted originally by the Canadian Army in October 1943 to tow the 6-pounder gun in reconnaissance regiments, infantry battalions, and motor battalions. Nonetheless, at the beginning of 1944, it was still undecided whether the T16 or the Loyd Carrier would be used to tow the 6-pounder. Deliveries of both were slow. By the middle of February, 229 T16 Carriers had been received out of a total requirement of 484. By April supplies were practically complete. The 3rd Canadian Infantry Division was issued with Universal Carriers as 6-pounder towers for the Normandy assault landing and continued to be equipped mainly with these. The T16 carrier was also adapted to carry the 4.2-inch mortar instead of towing it on a trailer as originally intended. One hundred suitably modified mortar

A Daimler II of either the 7th or 8th Canadian Reconnaissance Regiment. Note the 100-round magazine on the anti-aircraft mounted Bren gun. Photo courtesy Barry Beldam

US-built M3A1 White Scout Car, in Canadian service, it was designated Truck, 15cwt, Armoured. It's primary role was as an armoured personnel carrier in forward areas, but it could also be used as a command vehicle or as an amublance if modified. It was used by the Motor Infantry Battalion of the Armoured Brigades in the 4th and 5th Canadian Armoured Divisions and by the Assault Troop of an Armoured Reconnaissance Regiment, to name but a few examples of it's employment. Library & Archives Canada

Truck, 15cwt, Armoured (C15TA) whose primary role was as an armoured personnel carrier. This example benefits from the addition of chains fitted to the rear wheels to aid in traction on the icy surface. Library & Archives Canada e002505164

The Universal Scout Car. Along with the Caplad, this vehicle was limited to prototype stage only. Designed for a crew of three it featured a special transfer case which incorporated a reverse gear that allowed the full range of forward transmission speeds to be available in reverse. Library & Archives Canada

A Canadian-built Caplad (an acronym for Car, Armoured Personnel, Light Aid Detachment.) A logical development of the armoured truck, or maybe a wheeled version of the Universal Carrier. A multi-purpose armoured vehicle where the Canadian prototype was based on the General Motors Fox components. However, the war ended before they reached production status and the project was dropped. MilArt photo archive

carriers were available for issue by the end of April 1944. This use of the T16 Carrier was peculiar to the Canadian Army in North-West Europe and had some disadvantages. The bomb carrying capacity was limited and there was a lack of stowage space. The 4th Canadian Armoured Division recommended the use of M14 half-tracks as 4.2-inch mortar carriers to remedy this as well as to provide greater manoeuvrability. Due to the short supply of M14 half-tracks, it was decided that the T16 carrier would continue in use, but with modifications to overcome the faults outlined.

During early 1944, trials were made with new types of carrier. The Canadian Windsor Carrier was submitted for trial in February. This was similar in appearance to the T16 Carrier but was actually a lengthened version of the Canadian Universal Mark II, and had greater stowage capacity than either the T16 or Universal. The British Ministry of Supply placed an order for 5,000 to replace the Loyd Carrier as a 6-pounder tower. The Windsor would not replace the Universal Carrier. The Canadian Army decided to follow the British in this policy. Production of Windsor carriers was started in September 1944.

Armoured Personnel Carriers.

A number of different vehicles were considered for this comparatively new requirement. In September 1943, the requirements of the Canadian Army Overseas were about 600 armoured personnel carriers and 108 armoured ammunition carriers, 18 for each of six self-propelled artillery regiments. The vehicle which had been in use by the British as a personnel carrier was the American Car 4x4 Scout M3A1 (White Scout Car), but this was going out of production. The vehicles available were various American half-track carriers and the new Canadian Truck, 15-cwt, 4x4, Armoured (initially known as the "carrier wheeled", a name that was soon dropped in favour of the more descriptive "armoured 15-cwt truck"). As a long-term policy, the Canadian Army considered the Caplad, a Canadian 3-ton armoured lorry which

Morris Mark I, Light Reconnaissance Car. Used by the divisional reconnaissance regiments of the 1st, 2nd and 3rd Canadian Infantry Divisions, in the early years and in Italy/Sicily. Library & Archives Canada e002505504

would perform a wide variety of tasks. The Caplad (an acronym for Car, Armoured Personnel, Light Aid Detachment) was based on a UK design for a rear-engined, all purpose, armoured vehicle. It was generally simialr to the Fox I without the turret. The vehicle design was flawed by the 101-inch wheelbase, combined with a 90-inch width. Interior space was further lost to the sloping sides and driveshaft floor tunnels. It was finally decided that the Caplad could not satisfy any of the requirements originally laid out for the vehicle and the project was cancelled.

The policy finally adopted by the Canadian Army was to use the armoured 15-cwt truck as a personnel carrier and the American M14 15-cwt half-track truck as an ammunition carrier for the assault landing. In January 1944, as neither of these had yet been obtained by the Canadian Army, it was decided to issue White scout cars pending the availability of the armoured 15-cwt truck, of which 800 had been ordered. Production of the Canadian vehicle had been delayed by strikes at General Motors. In February, the continued delay resulted in a decision to adopt the British scale of issue for M14 half-tracks and White Scout Cars. This meant a greatly increased proportion of half-tracks. Both these vehicles were obtained through the British. By May, the Canadian Army had received sufficient vehicles to cover unit establishments. By this time the Canadian 15-cwt armoured vehicles were beginning to arrive, and 21 were issued to 3rd Canadian Infantry Division.

However, in the belief that demands would be for half-tracks rather than wheeled vehicles, steps were taken to ensure that only the original 800 armoured 15-cwt trucks were ordered from Canada. In June 1944, half-track trucks were officially adopted by the British and Canadian armies as 17-pounder towers, in lieu of field artillery tractors. Owing to the limited availability of half-tracks this policy was slow in coming into effect.

Kangaroos

Two heavier vehicles were developed by the Canadian Army in response to special needs. For Operation TOTALIZE (7-9 August 1944), the 72 Priests (105-mm SP guns) used by the three field regiments of 3rd Canadian Infantry Division for the assault landing in June, had their guns removed and were employed by the 4th Canadian Infantry Brigade as armoured personnel carriers. This was the first time heavily armoured carriers had been used to carry infantry

The Fox Armoured Car, a Canadian-built copy of the British Humber Mark III. Disliked mostly for its high profile it was limited to reconnaissance regiments and for training armoured car regiments in Britain before being replaced by the Staghound. Library & Archives Canada PA145009

forward into battle. In this new role the Priests came to be known as Kangaroos and continued to be used for several months as an informal squadron. Towards the end of the year it was reported that, *"Reduction in infantry casualties following experimental use of Priests stripped of armament and used as armoured personnel carriers has resulted in a 21st Army Group decision to provide such facilities in the form of armoured personnel carrier regiments on the scale of one per army"*. In accordance with this policy, 1st Canadian Armoured Personnel Carrier Regiment was formed in First Canadian Army. Meanwhile, a start had been made in converting a number of Ram tanks into armoured personnel and ammunition carriers. The new unit had an establishment of 106 armoured personnel carriers. At the end of November 1944, 66 of the Ram carriers were held by First Canadian Army.

Wheeled Armoured Fighting Vehicles

Wheeled Armoured Fighting Vehicles (AFV) used by the Canadian Army Overseas were almost all of Canadian manufacture and were of three main types. The first of these was the General Motors Otter, a light reconnaissance car. This vehicle was first received from Canada in the spring of 1942. The second type was the Canadian Lynx. This was a two man scout car, outwardly very similar to the British Daimler Dingo scout car. Built by Ford Motor Company Canada, it boasted a rear-engine, four-wheel drive chassis. It was a good deal more powerful than the British scout car, having a Ford V8 engine installed, but it lacked the advantages of the Daimler's independent suspension and Wilson transmission. Armed with a single Bren light machine-gun and a Boys rifle, it was used throughout the Canadian Armoured Corps and by Royal Canadian Corps of Signals units attached to armoured formations. The third Canadian wheeled AFV was the General Motors Fox armoured car. This vehicle carried a heavy machine-gun armament consisting of a .50 calibre heavy machine-gun, a .303 calibre Bren

US-built GMC Motor, Carriage, 75-mm Gun. Used by the Heavy Troop, in each of the four fighting squadrons of the 1st Armoured Car Regiment (Royal Canadian Dragoons) while in the Italian theatre of operations. Library & Archives Canada e002505168

light machine-gun and a .30 calibre medium machine-gun Browning, but no cannon. It was a four-wheel-drive vehicle with a crew of four. It was built on a Humber-pattern hull and turret mounted on a General Motors chassis.

As of 1 December 1943, the Canadian Army Overseas held 256 Fox armoured cars, 534 Lynx scout cars, and 648 Otter light reconnaissance cars. Although a number of Canadian Otters were issued to Royal Canadian Army Service Corps units, this was an unusual occurance. The Otter was to be replaced by a Universal scout car, described as a "General Utility Project". In the meantime it was proposed to provide a modified Lynx from Canada. The Fox had also been found inadequate and it was to be replaced by the newer American T17E1 Staghound, (*see below*), 51 of which had already been received.

Armoured Cars.

In February 1942 the 5th Canadian Armoured Division held eight armoured cars against an establishment of 60. In addition, it possessed 11 Ironsides which were rated as "Improvised AFVs". The Ironside was a conversion of the basic commercial model of the Humber Super Snipe, to meet the need for armoured cars, resulting in the Humberette, otherwise known as the Ironside but more officially as the Humber Light Reconnaissance Car Mark I.

Policy at the beginning of 1944 was to use the American Staghound Armoured Car (T17E1) except in reconnaissance regiments where the Canadian Fox was used. Infantry divisional reconnaissance regiments needed a car within the class nine bridge classification of infantry divisions, which excluded the Staghound. Consideration was given to following British policy in using some light armoured cars instead of all heavy cars within the armoured car regiment. The difficulty was supply. While sufficient Staghounds would be available, there were only three light armoured cars available, the Fox, Daimler, and Humber. There were plenty of Fox armoured cars but their suitability was questionable. Enough Daimlers could be obtained in time to cover the requirements of infantry divisional reconnaissance regiments only. The decision was taken to use Daimlers in reconnaissance regiments and Staghounds in

Lynx II Scout Car. The 'N/S' markings indicate that this vehicle is 'non serviceable' and that it was probably retreived from a salvage depot for the photo. Library & Archives Canada PA183191

Humber Mark III Light Reconnaissance Car of the 7th Reconnaissance Regiment (17th Duke of York's Royal Canadian Hussars), the reconnaissance regiment of the 3rd Canadian Infantry Division, Normandy, August 1944. Library & Archives Canada PA171698

Staghound with an experimental setup of four rocket launchers on either side. The damage on the fender was caused by the first experimental launch. Subsequent photos in the series show greater damage with each launch. Library & Archives Canada

Humber Mark IV Scout Car. Compare the British original, with its 37-mm gun, to the Canadian copy. Library & Archives Canada PA201348

US-built M8 (Greyhound) Armoured Car. This six-wheeled vehicle saw limited use in Italy and in Northwest Europe. The example shown here is from Headquarters Squadron, 1st Armoured Car Regiment (Royal Canadian Dragoons). Courtesy Royal Canadian Dragoons Museum

Otter Mark I, Car, Light Reconnaissance. This type of car was used by the infantry divisional reconnaissance regiments, by certain Field Squadrons/Companies of the Royal Canadian Engineers as well as by units of the Royal Canadian Army Service Corps. Library & Archives Canada e00205516

General Motors' 15-cwt Armoured Truck. Used primarily as an armoured personnel carrier in forward areas it was also converted to ammunition carrier and ambulance roles. MIlart photo archives.

GM Armoured ambulance. This was another project that never got beyond pilot stage. The war's end cancelled any demand for the vehicle. Milart photo archives

all other units. By the middle of February, 128 Staghounds had been received against an establishment of 129, although these did not include any command or control vehicles for which there was a requirement for a further 20. Some of these special types became available in April while the rest were covered by converting White scout cars.

The 18th Canadian Armoured Car Regiment continued to be equipped entirely with Staghounds, but they were replaced in formation headquarters with lighter cars as available. 18th Canadian Armoured Car regiment reported the Staghound to be a magnificent vehicle. In a classic example of the privileges of rank, Lieutenant-General Guy Simmonds, General

Otter I - Car, Light Reconnaissance. This car is in use by the 4th Reconnaissance Regiment (4th Princess Louise Dragoon Guards), the divisional reconnaissance regiment of the Canadian 1st Infantry Division, somewhere in Italy. It was also used by Field Squadrons /Companies, RCE for short special recces and by Bridge Companies, M.T. Companies and Transport Platoons, RCASC for AA protection of convoys. Armament consisted of one Bren LMG and one Boys Anti-Tank Rifle in hull when not fitted with wireless. Library & Archives Canada PA201349

Officer Commanding, 2th Canadian Corps appropriated two Staghounds as 'chargers'. Simmonds used these for personal reconnaissance and to visit units under his command.

The expected supply of Daimler armoured cars did not materialize and, in February, Humber IV cars were ordered instead. The Canadian Army obtained 56 of these at once, enough to cover establishment, and reserves followed soon after. In July, it was decided to re-equip the 7th and 8th Canadian Reconnaissance Regiments with Daimlers as they became available. By the end of November the conversion was complete. Humbers were still being used in other units. In October, 21st Army Group requested that all Fox armoured cars in the United Kingdom be held available for internal security purposes in North-west Europe. The Canadian Army in the United Kingdom held about 200.

In Italy, the 1st Canadian Armoured Car Regiment held 14 Staghounds and eight half-tracks against an establishment of 66 armoured cars. Daimlers were used to cover the rest of their requirements although future policy was to equip them entirely with Staghounds.

Late in the war the American M8 Reconnaissance car, known as the Greyhound, was issued in limited numbers to some Engineer units and, at the least, as Regimental Headquarters vehicles of the 1st Canadian Armoured Car Regiment (Royal Canadian Dragoons). The Greyhound was one of the few 6-wheeled vehicles used by the Canadians and suffered from weak floorboard protection against land mines. Nonetheless, in Italy, it gained an enviable reputation within reconnaissance units. In the immediate post-war period, Lt-General Simmonds requested that a Greyhound be made available to him as a personal vehicle but all of these had been returned to British control.

Lynx Mark II Scout Car. This type of car was used for short range reconnaissance by armoured car regiments, armoured reconnaissance regiments, armoured regiments, reconnaissance regiments headquarters of an armoured division and by Royal Canadian Corps of Signals in armoured formations. Milart photo archives.

Morris Mark II, Light Reconnaissance Car. Used by the divisional reconnaissance regiments of the 1st, 2nd and 3rd Canadian Infantry Divisions. Milart photo archives.

Scout Cars

Light reconnaissance cars (formerly termed armoured reconnaissance cars), of which a divisional reconnaissance battalion had an establishment of 45, were to all intents and purposes entirely lacking through 1940. During the week ending 19 April 1941, nine "Beaverette"

US-built GMC, DUKW 353, 2fi-ton, 6x6 Amphibian. Library & Archives Canada e002505162

armoured cars were issued to 4th Canadian Reconnaissance Battalion (1st Canadian Infantry Division) and the same number to 8th Canadian Reconnaissance Battalion (2nd Canadian Infantry Division). These machines, consisting of a light armoured body on a standard 14-horsepower commercial chassis, with two-wheel drive, were however considered very inferior except for training purposes. In fact, after eight Beaverettes were transferred to the 1st Canadian Army Tank Brigade in August 1942 their war diary referred to them as *"fast, dangerous, near-useless vehicles."*

By February 1942, the position of the Canadian Army Overseas remained unsatisfactory. The establishment of a divisional reconnaissance battalion called for 45 reconnaissance cars but the three Canadian infantry divisions possessed only 17, 13 and 14 cars respectively, and these were the inferior Beaverette. At this time the 5th Canadian Armoured Division held 11 Ironsides,

The problem of supply to a great extent governed Canadian policy regarding scout cars during 1944. The intention of replacing the Lynx I with the Lynx II was originally abandoned in December 1943 when it appeared that the Lynx II would not be ready in time. The War Office was asked to supply Humber I scout cars instead and the first were released to the Canadian Army in January 1944. But the estimated number available in time for operations was 150 out of a requirement of 188. Accordingly it was decided to go back to the Lynx I in 18th Canadian Armoured Car Regiment which required 65, and keep the Humbers to cover all other requirements. The Canadian Army was, finally, completely equipped on this basis early in April 1944

The main problem was to keep 18th Canadian Armoured Car Regiment equipped with scout cars. In June 1944, they were reported as satisfied with the Lynx I. But the wastage rate with these vehicles proved to be high. Although 75 reserves were held against an establishment of 65, by the middle of September it was estimated that these would be used up entirely in another six weeks. Further, it would take too long to overhaul these for operations and there was no prospect of being able to equip the armoured car regiment with Humbers. Accordingly it was decided to use the Lynx II, a number of which had now been shipped on the British order. The first of these had come off production in April and at that time it was felt that the Canadian army would have no use for them. The British intended to use them only in the Mediterranean or India, and the Canadian Army felt bound to follow British policy. The pro-

Two common amphibian vehicles were the Terrapin (left) and the LVT (Landing Vehicle Tracked) "Buffalo" Mark IV (right). There were two types of LVTs used, the LVT Mark II, which was a rear-engined model with a central hold, and the LVT Mark IV (pictured), which had a front-mounted engine and a hinged ramp that enabled it to transport light vehicles and guns in addition to the more usual payloads of troops or stores. Library & Archives Canada

Terrapin Mark I - manufactured by Morris-Commercial Motors in 1944, the Terrapin was a rival to the US-built DUKW. 5th Assault Regiment, 79th Armoured Division, were a user of this vehicle. Library & Archives Canada e002505440

cess of re-equipping 18th Canadian Armoured Car Regiment was started in November 1944.

Towards the end of August 1944, there was a growing need for scout cars for battalion commanders and for liaison officers of infantry brigades. Some Daimler scout cars began to be issued to the Canadian Army in August, but both these and the Humbers were still in short supply. Supplies were, however, always sufficient to cover establishment, and at the end of November there even existed a surplus of these vehicles.

Amphibian Vehicles

A particularly interesting development in 1944 was the number of amphibian vehicles used by the Canadian forces at various times. Mention has already been made of the amphib-

Vehicles belonging to British 8th Army Troops and to 1st Canadian Corps Troops, junction of the Liri and Gari rivers, Italy, 21 May 1944. The jeep in the immediate foreground centre belongs to Headquarters, I Canadian Corps. Library & Archives Canada PA151180

ian DD tanks used for the D-Day assault landing in June 1944. There were no other amphibian vehicles being used specifically by Canadians on this occasion, but considerable use was made by the British beach groups of DUKWs (The letters in the name DUKW identify the vehicle as; 'D' for 1942, 'U' for utility (amphibian), 'K' for front wheel drive, and 'W' to indicate two rear-driving axles) in landing stores and evacuating casualties. These were American amphibian 2½ ton trucks which were loaded at sea from the ships and carried the stores to the beaches and inland to supply dumps. The Canadian forces in Italy also made use of DUKWs in river crossings, both for reconnaissance purposes and in operating a ferry service

Amphibian vehicles were used on one notable occasion by the Canadians. This was the assault landing across the Savojaards Flaat during the clearing of the Scheldt Estuary in October 1944, when 9th Canadian Infantry Brigade was transported in Buffaloes. Operated by the Royal Engineers, they were armoured tracked vehicles which carried the brigade across the water and, on landing, crossed a mud flat, a slope studded with pickets, a grassy stretch, a dyke and a seven-foot wide drainage ditch. Two types were used, one for men and stores, the other with a door at the rear for vehicles in which jeeps, carriers, and 6-pounder anti-tank guns were carried. On the same occasion, Terrapins were used as load carriers. These, the British equivalent of the DUKW, were unarmoured 8-wheeled amphibian lorries. On a hard surface the leading wheels were held clear of the road, only coming into use on uneven ground, ie: when climbing out of a river. It had two engines, arranged to drive the wheels on either side so that one steered by slowing down one engine and causing the vehicle to skid round. It had two pro-

General Motors' Truck, Heavy Utility, Personnel - used as a light vehicle for transporting personnel in small parties, also employed within units as a unit Commanding Officer's office vehicle. It could also be used as a light ambulance. Milart photo archives

Transporter, 16-ton manufactured by the Four Wheel Drive Auto Company, shown here pulling a Trailers, 10-ton, GS and a Trailer, 6-ton, GS. Milart photo archives

pellers and a rudder for water-borne operation. Only 500 Terrapins were built.

No amphibian vehicles were on the establishment of the Canadian Army but one type was reported as being in the possession of First Canadian Army. This was the American M29 Weasel, a fully-tracked unarmoured vehicle intended as a load or personnel carrier primarily for swamps or on inland waters only. Examples were acquired by First Canadian Army in August 1944. Some Weasels were used during the clearing of the Scheldt Estuary. Weasels were also reported used for operations in Italy.

The amphibian Jeep (5-cwt Ford) was first considered by the Canadian Army in May

Sherman V, ARV Mark I, Armoured Recovery Vehicle shown loaded on a 40-ton trailer. The ARV was used by Armoured and Armoured Reconnaissance Regiments for recovery of bogged or bellied vehicles to a hull down position where unit fitters can work, as well as by RCOC/RCEME to recover casualties to a place out of range prior to being trailered to a repair area. Library & Archives Canada

Truck, 15-cwt, General Service. A workhorse of the army this truck, was used by all Arms and Services who appreciated its short turning circle and low silhouette. Milart photo archives

1943, but this, like other amphibian vehicles, was to be supplied by 21st Army Group. The first four were received in October 1944. A further vehicle being developed in Canada was the Willys Tracked Jeep, which had amphibian characteristics.

An Armoured Command Vehicle, High Power (Matador) 4x4. The Royal Canadian Corps of Signals at division headquarters used it to communicate between the division's main and rear headquarters and the main and rear headquarters of a Corps. Library & Archives Canada PA142060

Mechanical Transport

The vast majority of mechanical transport vehicles ("B" vehicles) came from Canadian sources, although certain specialist vehicles were British-made. At the end of October 1943, an inventory of British-made and Canadian-made "B" vehicles in the possession of the Canadian Army Overseas, showed no really large group of vehicles as being of British origin. The single largest group of British-made vehicles consisted of 197 Ford Commercial, General Service (GS), 3-ton Lorries with 158-inch wheelbase.

A few types of "B" Vehicles were supplied as a matter of policy from British sources. Several specialist vehicles for which there was a very small Canadian requirement were supplied in this way. One was the 3-ton Troop Carrying Lorry for which the British Bedford TCV was used. Production of a Canadian vehicle with the same capacity was being considered.

In September 1943 there were, in England, approximately 17,000 Canadian vehicles in crates awaiting assembly. Civilian assembly facilities were inadequate to maintain a sufficient flow of completed vehicles to the Canadian Army Overseas. In order to augment the rate of assembly a Canadian Equipment Assembly Company was formed; this was soon expanded into No. 1 Canadian Equipment Assembly Unit, comprising a headquarters and six companies. Up to early January 1944, this unit had assembled 3,423 vehicles, of which 1,902 were delivered complete, and 1,521 were awaiting final completion owing to shortages of parts. No. 1 Canadian Equipment Assembly Unit and the civilian firms under contract with CMHQ assembled approximately half of the 17,000 vehicles held by the Canadian Army by the end of 1943.

As late as 1944, the arrangement of acquiring a certain number of specialist vehicles from British sources was still being followed to a considerable extent. At the same time, the problem of specialist vehicles was being mitigated to some extent by the policy of modification of Canadian-made types in the United Kingdom. The system followed was to produce a minimum of basic types, and to then modify these overseas to meet specific technical requirements.

Truck, 3-ton, General Service, with early-style Type 11 cab. This type of truck was used generally by all Arms and Services of the Canadian Army as a standard load carrier. It was adaptable to special roles such as mobile kitchen, canteen, surgery, stores, etc.. by means of removable kits. Milart photo archives

Truck, 15-cwt, Fitted For Wireless (FFW). Manned by unit signallers, it was used mainly for wireless communication with Brigade and Divisional headquarters. It was employed wherever the Wireless Set No. 19 was suitable. Milart photo archives

One type of "B" vehicle which had appeared in the Canadian Army Overseas during 1943 deserves perhaps more than a passing mention. This was the American car 5-cwt 4x4, familiarly known as the 'Jeep'. This little vehicle had remarkable cross-country and hill-climbing performance, and had become very popular. All reports from Sicily emphasized the desire

Amphibious, 4x4 /-ton Jeep being demonstrated to Canadians in Iceland. Library & Archives Canada e002505505

of units in the field in the Mediterranean area to acquire as many of these very useful vehicles as possible. On 22 Sepember 1943, Lt.-Col. A.F.B. Knight, 1st Canadian Infantry Division, stated that formations and units of the Division in Sicily greatly preferred jeeps to carriers for the conditions under which they were operating. It was also noted that supplies of the popular Jeep were not always able to keep pace with increased demands. In May, First Canadian Army held a surplus of about 200, whereas at the end of November, with increased total holdings, there was a deficiency of nearly 300.

At the end of 1943 there was a shortage of 15-cwt wireless vehicles which led to the employment of Heavy Utility Trucks in a role for which the 15-cwt was preferred.

Motorcycles

The decision to change from the heavy Harley-Davidson motorcycle of US manufacture supplied from Canada to the light motorcycle using the British Norton and Matchless was taken in the summer of 1943. The changeover was completed early in 1944. First Canadian Army continued to hold a few Harley-Davidson machines as well as a few other British makes, but the great majority of the motorcycles held were Norton and Matchless. In November 1944, it was decided to use the Harley-Davidson once more in some Royal Canadian Army Service Corps units, as well as for Provost Corps units in Army Troops.

Epilogue

With the surrender of Germany's forces in May 1945 the war in North-West Europe came to an end. The Canadian army, and Canada as a whole, could look back over the previous six years with well-earned satisfaction at their accomplishments. From an uncertain beginning, the army had matured into a respected fighting force capable of meeting any challenge. With a small core of professional officers, combined with a nucleus of soldiers and noncommissioned officers provided by the prewar Permanent Force, the Canadian army grew into the Western Allies' third largest army. This army was manned originally by members of the militia, followed by volunteers and, in response to a pressing need for reinforcements, a number of conscripted soldiers. Canada sent five divisions and two independent armoured brigades to Europe. In addition, Canada staffed a headquarters in London, as well as an Army and two Corps headquarters in the field. In Great Britain, the Canadian Army staffed its own ordnance

A more realistic portrayal of a jeep in a combat zone. The jeep's ability to go everywhere, coupled with its ease of operation, made it a popular vehicle with soldiers of the Allied armies. This jeep is in use as a two-berth Ambulance Car of No. 2 Light Field Ambulance, RCAMC, 1st Canadian Armoured Brigade. Library & Archives Canada PA205328

Car, 5cwt (Jeep, 5cwt, 4x4) An example of a Canadian contract jeep. Note the lifting rings on the front bumper as well as the formation sign of the Overseas Canadian Training School. Library & Archives Canada e002344119

depots, reinforcement units and training schools. The Canadian army provided soldiers to every theatre of operations and, in some instances, provided large numbers of junior officers to the British army to meet that army's needs under the Canloan programme. Canada also provided a number of officers and senior NCOs to the North African campaign for battle experi-

Ford, Pygmy, 5-cwt, 4x4. Early version of the ubiquitous jeep, whose primary role was that of a personnel carrier, but it could be fitted with stretchers or used to tow light anti-tank and anti-aircraft weapons as well as to transport light cargo. Library & Archives Canada PA140161

Model M 20 (B) BSA Motorcycle. Motorcycles were used to provide rapid transportation for despatch riders and other personnel whose duties required them to move between various points, such as the Canadian Provost Corps and personnel of the Royal Canadian Army Service Corps. Additionally, all officers were trained in motorcycle riding. Canadia Library & Archives Canada PA134479

ence. Canadians also served as part of the bi-national Special Service Force under American command.

In Canada, the army raised three additional divisions, one of which was ear-marked for the Pacific theatre. This home army was later reorganised into the Pacific and Atlantic

Norton Motorcycle, Solo, Light, of the Overseas Canadian Training School Library & Archives Canada e002344122

Commands. The army provided recruitment services, as well as basic and advanced training for all arms. These units also drew on Canadian industry for the same equipment then being provided to the army overseas.

In 1939, Canada boasted little in the way of motorised transport but by 1945 had produced almost a million Canadian military pattern (CMP) trucks, and a large quantity of modified conventional pattern (MCP) trucks, and supplied these worldwide to virtually every allied nation.

By war's end, Canada boasted a small arms industry that saw every type and pattern of individual weapon manufactured in Canada, including rifles, pistols, machine guns, grenades, mortars and anti-tank weapons, as well as the ammunition for these. Larger weapons systems were also produced by Canadian factories including 6-pounder and 25-pounder guns as well as the carriage for 5.5-inch artillery.

The army which, prewar, relied on crude plywood mockups as substitutes for mechanised vehicles fielded two armoured divisions and two independent armoured brigades against the enemy. Canadian industry, in spite of their initial lack of expertise, managed to produce a respectable number of tanks and made many credible design suggestions which were later reflected in the more numerous American cruiser tanks. Light armoured vehicles such as the Universal and Windsor carriers equipped Canadian forces, as well as its allies through the Mutual Aid programme. Canada also designed and produced a number of scout and reconnaissance vehicles.

The army, which was relegated to using obsolescent artillery in 1939, ended the war with a modern artillery corps. More important, Canada had acquired the ability to design and manufacture artillery equal to that found anywhere else. Combining our newly found talent as manufacturers of armoured fighting vehicles with our ability to design and manufacture artillery led to the development of the Sexton 25-pounder self-propelled gun which became a mainstay of the British and Canadian armies.

Canadian industrial inventiveness resulted in improvements to optics, small arms, flame weapons and much more, while, in the field, Canadians developed new equipment, such as the

US-built Harley-Davidson WLC. In Canadian service this was designated Motorcycle, Solo, Heavy.
Library & Archives Canada e002343942

Land Mattress rocket launcher and the Kangaroo armoured personnel carrier, to meet the operational needs of the fighting soldier.

The Canadian Army, which saw almost 260,000 men and women serve in Northwest Europe, and more whose service was limited to North America, was soon reduced, by 1947 to less than 32,000. The amount of equipment available was staggering and much of it was not brought back to Canada. Large quantities of military goods were sold to European nations who, after years of occupation, were rebuilding their armies. As these countries could not afford to pay for the Canadian surplus, Canada lent them the necessary money and then forgave the loans. In addition, thousands of trucks were donated to the United Nations High Commission for Refugees. Back in Canada, the industries which had been met the challenge of manufacturing war materiel turned their attention to producing consumer goods to meet the needs of the returning soldiers as they resumed their pre-war life. In most cases this involved laying off the hundreds of thousands of female workers and trying to integrate the demobilized soldiers back into industry. A problem was created when a large proportion of former agrarian workers chose to stay in high-paying factory jobs rather than return to the family farm, but subsidies helped return them to the fields.

For many industries the return to peace-time conditions was welcome and relatively event-free. For some, however, especially those established by the government as 'Crown corporations' there was no transition and these businesses were either merged with other viable 'Crowns' or shut down completely. This was especially true for those industries which had specialised in purely military goods and for which there was no demand in the post-war era.

Canada would never again see its army as well trained, well equipped or as large as it was in May 1945.